Information Technology and Cyberspace

Other titles in the series:

Reproductive Technology:
Towards a Theology of Procreative Stewardship
by Brent Waters

Forthcoming titles include:

Punishment
by Christopher Jones

Human Genetics
by Robert Song

Euthanasia
by Nigel Biggar

Information Technology and Cyberspace

Extra-connected Living

DAVID PULLINGER

THE
PILGRIM
PRESS
Cleveland

Published in the USA (only) by
The Pilgrim Press
700 Prospect Avenue East
Cleveland, Ohio 44115–1100
pilgrimpress.com

Originally published in 2001 by
Darton, Longman and Todd Ltd
1 Spencer Court
140–142 Wandsworth High Street
London SW18 4JJ

ISBN 0–8298–1464–7

06 05 04 03 02 01   5 4 3 2 1

A catalogue record for this book is available from the Library of Congress

Designed by Sandie Boccacci

Printed and bound in Great Britain

Contents

Acknowledgements

IN WRITING I have been helped by a number of conversations, some taking place over years, some in response to my lectures, others over breakfast in hotels, and still others when I rang people and asked for guidance on some particular point. The latter have included David Leal, who introduced me to the work of Stanley W. Hauerwas and to specific passages in Barth, and David Ford, who introduced me to Paul Ricoeur's writing and suggested I read his book *Self and Salvation*, which I did with great profit.

The approach of 'living from below' in relations of power emerged when I was teaching students in the Faculty of Divinity, University of Glasgow, and the Department of Computer Science, Heriot-Watt University, Edinburgh, where one semester I lectured in both universities alternately and sought to say something that made sense to both.

Part of Chapter 5 was originally researched for the Church of England's Board of Social Responsibility report *Cybernauts Awake!* Some of my discussions with colleagues there emerged into this book, rather than that.

Darton, Longman and Todd persuaded an eager catalyst but reluctant author to write – thanks to the editor Katie Worrall for her encouragement.

For the ongoing exploration of what it means to live as a Christian in a technological culture, I am appreciative of the support of St Mary Islington, London, and the clergy there, especially Rosy Fairhurst, and the Bible study group with whom I meet weekly: Fiona Green, Ruth Ogier, Ruth Thomas, Katharine Maycock, Rowan James, Arif Mohamed, Helen Taylor, Claire Weldin, Jonathan Gebbie and Emma Lloyd. Two members of the congregation, Michael and Harriet Maunsell, kindly filled me in a little on the legal aspects. The opportunity to rehearse and conduct John Tavener's 'God is with us' during the period of writing helped enormously in those moments when writing seemed an empty activity and I would rather be talking and helping people directly.

My greatest gratitude is for my sister-in-law Debbie Pullinger, who started as a research assistant (paid at market rates) but then volunteered to help at the final stages and shared that bubble of separate existence that comes as a book is finally put together.

chapter one

Introduction

> Are there any ethics on the Internet? Are ethics and information technology (IT) completely unrelated? Does 'ethics' imply simply 'stopping things' – taking a negative attitude to new developments?
>
> The Internet is:
> - an information store;
> - a means of communication and relationships;
> - a means of 'play', exploring our identity;
> - a new way of conducting business and politics.
>
> How can we help to ensure that human lives flourish in the changing 'social spaces' of the world of cyberspace? By taking a contextual approach to IT, we will be able to see more clearly the ethical questions that are raised. Technology itself, the T of IT, is neither good, nor bad, nor neutral. Christians have the same struggle as everyone else: this book will seek to 'position' the debate in people's lives, rather than provide answers to the questions.

'ETHICS? THERE AREN'T ANY' was the most common response to my statement that I was writing a book on the Internet and ethics. After a while, it became apparent that my friends were actually giving two different types of response. The first type said that there was no connection: that ethics and information technology belong in two separate fields and do not touch or relate to each other. For those people, simply listing the areas I intended to cover was enough to convince them to change their minds. This was one of the key goals of writing – to point to where there are in fact ethical issues.

The other type of response was given by people who already appreciated some of the problems – invasion of privacy and access to pornography were issues familiar to many – but they felt that since no one can control anything on Internet, 'no ethics are possible'. A second goal is therefore to examine ways in which ethics can be derived, although it may not, perhaps, be the prescriptive ethical framework that might be expected from a Christian book on ethics.

There was a third type of response, a kind of 'downsayer' from those who interpret the word 'ethics' as 'stopping things', or 'stopping people from doing things'. This line seemed to be particularly prevalent amongst some cyberspace developers (we will define cyberspace in a moment). They believe that they are helping to develop a new frontier, a new land without any of the problems of the material world – a new land of data and information, of electronic business and virtual reality, without old historic laws and national boundaries, without geography and without time. Any hint that we should discuss the possible impact of cyberspace on people's lives and on business is seen as anathema. Instead, they thought I should be promoting and marketing cyberspace as the future: as the solution to every-one's problems and as a salvation for the world. They seemed to suspect that I might give reasons for not using the Internet and hence foreclose the possibilities of this future.

As someone who developed some early large information sites on Internet, spent over twenty years working in electronic publishing and is now involved in e-commerce, I certainly do not write from any position directly opposed to its develop-ment. But to ask a question is not to be oppositional; it simply means that we do not accept blindly and in totality the hype presented by techno-evangelists. It seems reasonable to ask if their announcements that we already see 'global links on a scale unparalled in human history, tearing down petty, parochial interests while creating a global culture'[1] is either true or even desirable. Indeed some have seen the struggle for identity in smaller 'nations' as part of a response to globalisation.[2]

To think of ethics merely as a process of 'stopping things' is erroneous. Ethics are the principles of behaviour which underlie the decisions and behaviour that create good and fair societies in which all human beings flourish. That means, to take a couple of practical examples, that electricity utilities cannot choose to switch off power, just because they happen to have a few technical problems that day and it would be the easiest way to solve them; and companies who dislike a tax regime cannot withhold fuel in order to bring a country to a standstill. The responsible provision of these infrastructures to citizens is part of ethics. The Internet may soon be considered just such a utility.

It is undeniable that the Internet is and will be changing patterns of relating. The kinds of questions I therefore want to raise are about the relationships and communities fostered by communications technology: do communities and relationships over Web form extra links between people or replace them? Do Internet communities form a new kind of grouping that requires new governance? Who is my neighbour: the person living in my street or the one I meet somewhere in cyberspace? It may not be ethics, as many understand it, but the articulation of these questions lies within ethical discourse. To use the word 'ethics' in this context is to consider how best life can be structured in order that all may flourish.

Social spaces

One of the main problems inherent in this task is the pervasiveness of information technology – quite simply, what does not have a 'chip' in it? Living in the Western world means buying, using and interacting with tools and equipment that have information technology inside, whether it is to wash clothes, make a phone call or to buy books from another part of the world to be delivered tomorrow. We can interact with our technology through keyboards, moving pointers, speaking and hearing, and using special equipment, by direct body movement and pressure – and that by no means exhausts the list. There are few areas of life in which information technology does not play some part. Books giving overviews cover law, politics, economics, organisation, work and employment, security, privacy, professional conduct, media, artificial intelligence, education, health, government, defence, crime, robots, regulation, patents (the contents of just one of my library shelves). Fully to address even one of these areas in relation to information technology would be beyond the scope of this book. However, there are a number of themes that keep popping up in different places, irrespective of where you begin. Probably the two most conspicuous ones, especially in US writing on the subjects, are privacy and individual autonomy. These are closely followed by the breakdown – and reconstruction – of community life; the 'virtualisation' of living; the loss of face-to-face interaction; the question of who our neighbours are; and what political authority means now that we so freely range over borders on the Internet (buying goods, for example) with little apparent control. All these concern the social dimensions of our life and are the main ones that I concentrate upon in this book. They are not to do with 'chips' or the technical aspects, but with the direct use we make of the tools of information technology in informing ourselves, communicating with others, and transacting business. We are not talking about a technology that we simply use as a tool, but one that changes social structures: how we accomplish our endeavours, how we make money and survive, and how we relate to each other.

How life can flourish, as the objective of ethics, differs according to whether we are considering the Internet as an information store, or as a means of supporting communication and relationships, or as a means of 'play' and exploring our identity, or as a new way of conducting business and politics. These areas occupy different social spaces – and there are different social actors in each.

If we think of it as an information store, then the types of questions that arise are about the accuracy of the information, about the processing of that information and about its use and value. These considerations lead to further questions about how we are informed and gain knowledge – of particular interest to business trying to fill the competency gaps they have in their organisations – and the ability of people to deal with increasing amounts of information, and so to questions of human attention.

If we consider the Internet as a communications system, with its email, chatrooms, video mail, etc., then another set of questions are raised. Here the discourse is about social structure – friendship patterns, the nature of community and whether online communities can be considered 'real' in any sense, who your neighbour is when people you relate to most days are located on the other side of the world.

For some people the Internet is primarily a means of 'play', where they can adopt different personas and so explore what it would mean to be someone else – whether they do this for fun, for psychotherapeutic reasons, or for enticement. This space raises questions of identity and leads to the wider consideration of how we define both our identity and our boundaries in a postmodern, technological society: who am I when I can be 'this' to one person online and 'that' to another face to face?

Finally, the Internet occupies national and trans-national social spaces. Here, people are considering whether cyberspace might be a way of laying claim to ideals of democracy. But there are other, perhaps more important questions to do with the way politics is used to shape cyberspace and its integration into the rest of life.

Each of the these social spaces raises different kinds of ethical questions, and so I will consider these social spaces separately, in turn, through the central chapters of the book.

Contextualising IT

The pervasiveness of information technology, even within the social spaces we described, brings with it another kind of problem. How do you have a discourse about it? I will adopt the approach of looking at the context in which it is introduced and used: of *contextualising* IT, rather than examining it as a separate technical development that imposes itself, as it were, from outside into our lives. As we appropriate it, we change our social patterns of living – how these are reshaped and the questions that are raised are what I choose to examine in this book.

We therefore have to be able to look at IT in context from a number of perspectives. IT has a politico-economic history – why do governments and organisations continue to invest so heavily in this, rather than alternatives? There has been technical development to get us to this point, but it will not stand still, so IT is located in a continuing stream of change. Digitisation that is at the root of these technical developments will continue to be applied in increasingly diverse areas of life. The Internet, the networks that connect computers together, also has a history and a different context from the purely technical one that supports IT. It is also located within the story of human attempts to transcend time and space.

To most of us, IT appears primarily as personal tools or services or, at work, as corporate infrastructure to enable us to do our jobs. By looking at these contexts, I hope that we will see that there are directions to the developments (and they don't all go in the same direction), and so we will be in a better position to choose whether or not to support them. The next chapter in the book goes through some of these perspectives.

T is for technology

IT is a technology. Many discussions begin with the assertion of the neutrality of technology – it's neutral, the proponents of the idea suggest, and we can use it in good ways or bad ways.

I do not believe that it is morally neutral. It is, for example, developed by particular interests for purposes that benefit them – and often benefit them over and against others. Consumers, for example, will pay for the digital television revolution. However, 'In the words of an insider, "There is no actual need for digital television. The need is for private information that would not otherwise be available."'[3] The primary goal of the digital revolution is based on obtaining a clearer picture *of* the viewer rather than *for* the viewer.

A second line I want to refute is one that favours waiting to look at the consequences of any new technology. If it turns out to have negative consequences, we can then agree that it should not continue. The trouble with the consequentialist approach is that you only ever know after the event. It is possible now to display films onto the windscreens of cars so that the driver does not get bored on long journeys, but few would recommend the approach of consequentialism for deciding whether or not to do it! Of course that is a case whose results we can predict fairly accurately, just as we already can equate the average speed of motors with a specific death and injury rate. It gets much harder to make the assessement with new ideas and so we need a process that is not based on experience to help us decide. This is where the study of ethics is valuable.

The starting point for many is observation of the world, and in doing this people tend to come away with either optimistic or pessimistic views. Those looking at technology itself tend to see the useful possibilities of its development, and not too many disadvantages, so they are optimistic about the next new introduction.[4] Those looking at social equality and relations of power tend to see more division in society, greater wealth disparities, regions of the world being excluded from the benefits – and come away pessimistic.[5] The two groups thus approach technology with spectacles that are either optimistic or pessimistic – each with good cause to do so. The question is, where do you stand in order to take off those spectacles?

The postmodern would argue that there is no place to stand, that we all stand in our own places and see the world from there with our own views – there is no possibility of finding a place from which to view the world in such a way that others could agree they would see the same. In their terminology, there is no meta-narrative. We do not have to debate the truth of this in order to see that many share some hopes and dreams in common; and that the concerns raised by IT keep recurring – some for over 2,500 years. There might or might not be a shared meta-narrative in a pluralist society, but so much commonality of concern means we do not have to abandon the task of identifying at least possible narratives in which to share.

Because the T of IT emerges in most discussions on internet, Chapter 3 addresses this directly. In that chapter I also declare the colour of my own spectacles (through which my viewing is done) as Christian and so, depending on your view, how my view is distorted or clear. By being explicit about this, I am leaving you to draw alternative conclusions from the evidence I present. Some argue that technology and religion have nothing to say to each other, and some theologians of different religions have agreed. My own view is that we shape the world by what we choose to use and adopt; by what we accept silently and what we argue against – most obviously in a consumer culture of capitalism. That is where I am to be found, looking at the technology offered and deciding to accept with gratitude, adapt it with struggle or to oppose it. I shall argue that this position is one that Christians in Western culture can adopt. There is, however, no clear theology of technology, and technology (in distinction to science) has been a much neglected area of Christian understanding with a few notable exceptions.[6] This is not the book in which to attempt such a theology, but I have tried to set out my position in the context of the debates.

Why should Christians be interested in information technology?

Christians struggle just as much as anyone else in trying to make sense of the world. Questions of identity, community, what it means to be human, the structuring of life for flourishing and justice, come to those of faith and to those without. The Judaeo-Christian tradition has been particularly interested in the kinds of questions that are being thought through afresh as a result of cyberspace – the identity that derives from bearing the name of God's people, deciding who one's neighbour is and what abundant life is and how to pursue peace and justice. All these are terms pregnant with meaning. For a follower of Jesus, is it possible to think of identity separately from who Jesus claimed to be and apart from a relationship with him? Is it possible to use the word 'neighbour' and *not* think of the story of the 'Good Samaritan'? When flourishing life is discussed, thoughts turn to the claim that Jesus came that people might have abundant life. But as the social world changes, we have to ask whether these stories recorded in the Bible are sufficiently robust to stand inspection and give guidance for a way of living two millennia – and nearly the whole of technological development – later. Do the meanings of identity in Christ, the concept of 'neighbour' and the ideas contained in 'abundant life' have to be reinterpreted anew? Or does Jesus' teaching get so distant that there is a struggle to find any meaning in it that relates to current social patterns dependent on technology?

Why should those who are religious attend to information technology? The Church in its 'good works' has always been attentive to the needs of the excluded, the poor, those affected adversely by change that go unnoticed by the dominant forces in society. Commentators are already noting the exclusion of the African countries, for example, in access and use of Internet, and there are widespread predictions of a two-tier society, one with access and the other without, for reasons of literacy, technical awareness, or money.

I don't, however, offer any quick and easy slogans. This is unashamedly an ideas and 'signposts' book, seeking to stimulate readers into making connections and pointing the way by encouraging their own exploration and drawing on their particular experience, expertise and insights. I have sought to position the debate on IT in our lives as we experience them, so that this task can be carried out.

Summary

In summary, my goals for this book are the following:

- to remove the surprise element many experience with the eruption of cyberspace into everyday life;
- to map out some areas and show how they relate one to another;
- to enable non-technical people to engage with questions and feel they can both do so and should do so without being put down by those claiming to be more expert;
- to help Christians to engage with changing social structures in forming their own lives, in their churches and at work.

It is an exercise of exploration rather than easy answers, as I have said, and, as with the early mappers, I fear that some mistakes might be made – a mountain not recognised as a volcano that might explode without notice; an unidentified region being marked 'here be dragons' where none exist. Like a mapper, I have drawn boundaries, given opinions, and all in the hope that others will travel out and explore further.

Finally, continuing the analogy, the maps I have drawn are very much the result of my own personal explorations and voyages of discovery. I include, along with maps, first-hand accounts of some of the things I found there; readers will undoubtedly notice that throughout the book I draw much on my own personal narrative and experience. At the risk of appearing to be, somehow, less 'objective' or 'academic', I include these in the hope that readers may find it easier to relate the ideas to their own personal narrative, and, by even the smallest of incidents or ideas, be inspired and encouraged in their own voyage of exploration of the technological life-world.

chapter two

Locating Information Technology and Cyberspace

Here we sketch out the origins and some lines of development ('lineages') of information technology and cyberspace, which help to locate them in the present-day world and account for the varying views people have of them.

We look at four lineages, in their chronological order:
- digitisation
- scientific management
- the Internet
- virtualisation

How do they emphasise the social spaces for individuals, citizens, or consumers in Western society? What ethical and social issues are brought to light?

WHAT IS INFORMATION TECHNOLOGY (IT) and where has it come from? Until the World Wide Web, people had home computers and software (if they had them at all) only for word-processing, doing accounts and perhaps for a few educational or entertainment programs. Most use of computers for information technology was in the workplace. Then, in 1998 and 1999, there was an ostensible change with the birth of new 'dotcom' companies (named after their World Wide Web addresses that usually ended in .com). These offered new and attractive services for individual consumers who only needed to have a computer and a modem and then 'get online' to take advantage. Access to a large consumer market (rather than a purely business one) launched many dotcom businesses into the public spotlight, not least because pension stock was being invested in them and consumers wanted to know about the companies and their prospects. The flurry of newspaper stories and media attention in 1999 took stock prices high and then, with no immediate profits showing by 2000, low again. The momentary drama, however, should not divert our attention from the long story leading up to this point, nor from the long one stretching ahead into future developments. It is a story made up of many intertwining threads or lineages. This chapter examines some of these lineages, simply to show where information technology has come from and why people say such a startling variety of different things about it.

First we consider the technical aspect that makes it possible – digitisation – and therein distinguish the different streams within its development: information, communication, e-business and virtual reality (VR).

Second, to create an infrastructure as large as this requires investment. Why do organisations invest? What is their goal? We follow commentators who see IT lying on a historical line of increasing 'efficiency' that began with the introduction of 'scientific management'. Further along this line, IT facilitates the construction of large, global corporations based on information-monitoring and control, with centralised decision-making.

Third, global research and business require worldwide communications systems that are reliable and robust. The Internet is the technology for achieving this and we trace its history – noting that all the major changes for people as consumers stem from its introduction.

Fourth, many commentators see the history of communications technology in terms of human attempts to transcend the limitations of time and space and the physical body. From this slightly different perspective on history, we consider how information and communication technologies continue this process.

These four threads are not the only ones: one could follow

many others, for example the interaction of our technologies with our thinking about ourselves as human beings – and the use of a computing metaphor for our brains. Another might be technology as part of the continuing scientific exploration, as a tool in the process for gaining knowledge. Major commentators have seen the studies of genes and DNA in this way, observing that they are data-derived, and stored and studied by means of information technology.[1] I describe these four lineages, partly to illustrate the variety of different approaches that can be taken in a discussion of the ethical issues. The ones I have chosen also emphasise the social spaces for individuals living as citizens and consumers in Western society and underlie the ethical and social issues covered in later chapters.

Information technology could be defined as all the technical bits and pieces that go to make up the technology combined with its uses by organisations and individuals. But that is to overlook another important aspect: the ways in which it makes people think differently about what is possible and how they interact with the world. The motor car provides a good illustration of this.

A car is not only a means of personal transport, provided there are some roads to enable driving around, but also a whole way of living. The car helps define who our friends are, where and how we meet them, what we do with our leisure, where we shop; how far we live from parents, children and grandchildren. For certain people such as mothers with young children, and the disabled, it means being able to get out to places that would otherwise be inaccessible. We think of the car as an extension of ourselves and our capabilities. Moving beyond the personal, the road transport system underpins all aspects of Western society, from the provision of law and order and emergency services to the production and daily delivery of goods to every town and city. Both government and commerce think 'through the car'. For most in the Western world, living without the car system would be highly problematic, both practically and psychologically. This is just one example of an integrated technology changing the very way we look at and think about and interact with the world. This effect is also apparent in technologies for information and communication and here it is described by the word 'cyberspace' – the place, it is claimed, where new thinking occurs, primarily because there are no longer limits on space or time. We explore some of these ideas later in this chapter.

Digitisation – the key technical development

If anything can be digitised, it can be 'treated' and 'output' in that or another form. The same basic process can be used to improve my holiday snaps on computer, to change the digitised sound of a stick hitting a bit of rubber to generate a drum machine, to take the sound of the voice singing down a micro-phone and correct the pitch before it reaches the loudspeakers or even transform it to sound like a flute or strings, or to take a spoken sound and interpret it as a word, or to analyse movements of bodies to help athletes, and to retrain the neuro-muscular patterns of victims of accidents using exo-skeletal body interfaces. Anything that can be digitised can be manipu-lated by software programs and then output in an altered form. The underlying power of the information age lies in this.

Digitisation refers to the process of recording data as a series of numbers; of zeros and ones, or yes–noes or on–offs. Once recorded in this way, data can be searched. You might search for patterns, or gene sequences, or words in a document that don't match those in the computer's dictionary, or words on web pages – so you can find a particular type of website, or words in an email – to get a general picture of what people are saying, or any sequence of sixteen numbers – possibly a credit card number. The whole point of digitisation is to put data into a form that can be easily retrieved and then analysed or transformed. It can also be transferred to a completely different process: for example, DNA sequences, once digitised, can be put into a music program to see if it sounds 'tuneful', or into a drawing program to see if it looks 'pretty'.

The systematic collection of data that can be searched for patterns generates a huge increase in power of analysis and creates new implications for possible actions based on information gained. Genetic engineering, for example, relies on this capability. Such is the task of analysis that many scientists do not collect data from life, but work on the data others have collected and put into a databank. There is software that allows an individual's home computer to download (automatically) data collected from space signals and then analyse it to see if there are any non-standard patterns – thus engaging a huge resource for the analysis of continuous data. The same techniques of digitisation and analysis are used on every consumer's buying patterns and every motorist's speed on the highway (captured number-plate images and speed measuring-device output are digitised, analysed and linked automatically to databanks and printers of demands for payment). New applications for these techniques are being found all the time – one such, being considered, is the analysis of the patterns of behaviour of people in London underground rail stations in order to identify those likely to be on the verge of committing suicide under a train – with the output being sent to a warning device for staff or to a taped announcement.

Of course the digitised output need not match real life; it might be purely imaginary, as when, in an animation, the head of a person turns round 360 degrees, or the mouth of a bird is made to articulate in the manner of a human, or an imaginary landscape is depicted. This is invaluable in training operators for events that they will rarely or perhaps never experience in real life, and that we would not want to occur, such as accidents in nuclear power stations, or severe storms or icing affecting the handling of aircraft. Computer games, whether they practise skills in hand–eye co-ordination or in problem-solving, involve the same imaginary output. Here the situation might be in imitation of life or one of pure fantasy – dodging giant spiders, facing enemies designed to kill you with a moment's inattention, or avoiding glass spheres bouncing around the room you are 'in'. Here, the term 'virtual reality' is often applied, especially when the user is totally immersed by means of multiple inputs – in addition to receiving vision and sound the body might also be moved by a device.

The technical accomplishments in software, hardware and communication to achieve digitisation, and the subsequent manipulation of data with output back into the world have been immense – and, at the start of the millennium, they are only just beginning. What then is the argument that any significant change should not be attributed to technical factors but more properly to other causes?[2] To find and understand it, we examine what led to the systematic collection of information in industrial society.

Scientific management – the start of the information revolution?

Why do we have the kind of systematic collection of information in modern societies that we do? One clue can be found in the story of IT development itself.[3] It is suggested that there were three phases spanning the period from the 1950s to 1990s. During the first phase, seeking cost-effectiveness by process control and automation of clerical tasks was the primary characteristic. Focusing on the benefits of computers and system software produced the first computer-based information systems. The second phase was marked by an increase in productivity of means for developing IT. The demand for IT professionals outstripped the number available, so new software tools were developed to help program more quickly. Then in the 1980s and 1990s, making software and data available to non-specialists and opening up information systems to many different kinds of users, trained and untrained, marked the third phase. Interfaces were designed to be easier to use and more intuitive – largely through the introduction of windows and pointers and a 'mouse' with which to 'click'. Since this particular account was published, a fourth phase has got under way: this centres on connectivity, and we will address it a little later. This version of history clearly points to the primary source of IT being one of process control and automation of clerical tasks for the sake of cost-effectiveness. This is the 'efficiency' argument – that the origin of the information revolution lies not in technical innovation, but in the drive for efficiency in production.[4]

Scientific management, claim Robins and Webster, among others, initiated the information revolution as we know it, requiring 'the dual articulation of information/knowledge for efficient planning and control'. Frederick Winslow Taylor, the advocate of scientific management, took efficiency as his main objective and sought to apply engineering principles to industrial production. 'Workers should be relieved of the work of planning and all "brain work" should be centred in the factory's planning department.' This required two steps – first, information-gathering by monitoring what the workforce and machines were doing and, second, having control over the workforce so that what they did could be modified to increase efficiency. Surveillance (used in the sense of monitoring everything that was happening) and administration to support decision-making were the underpinning strategies. Of course, to begin with, this was accomplished with a redesign of organisation with direct management – there was no IT. Henry Ford was the first to incorporate the idea into a technical system, on the car production lines. As the history of IT development indicates, IT fitted in well to assist in monitoring activities and to provide planning and control for the sake of cost-efficiency. But even before then, around the 1920s, scientific management held out the promise of social wealth through the increase of productivity and economic growth.

Mass productivity requires mass consumption. How could the two be linked? It would be very inefficient to have production when consumers didn't want the product, and to have needs not met.[5] The answer was to apply the principles of scientific management to consumption. Sloan at General Motors in the 1920s introduced many of the ideas of modern marketing – used-car trade-ins, annual model changes and instalment selling – so as to integrate production and demand.[6] This was possible only with information: collecting it, aggregating it, analysing it and disseminating it to those who wanted to know.

Henry C. Link, an ardent advocate of scientific marketing, described the relation between early forms of information technology and the informational needs of business:

> The most highly developed technique for measuring buying behaviour is that made possible by the electric sorting and tabulating machines. These ingenious devices have made it possible to record and classify the behaviour of the buying public as well as the behaviour of those who serve that public on a scale heretofore impracticable. Whereas by ordinary methods, hundreds of transactions may be recorded, by this method thousands may be recorded with greater ease. Not only have comprehensive records been made possible but, what is more important, the deduction from those records of important summaries and significant facts have been made relatively easy. The technique developed by various merchants, with use of these devices ... is the quantitative study and analysis of human behaviour in the nth degree.[7]

Conversely, consumers also need to know about products, so information must flow to them through packaging, branding and advertising. They need media to find out what is available for purchase, what is good value and what is safe.

What do companies have to do to keep on growing? Prevailing wisdom says that they have to extend the reach of their market through the world, to extend the range of products by careful research and needs assessment, and to speed up the whole production–consumption cycle by removing as many layers of planning and thinking as possible. 'The evolution of corporate structure can be seen as entailing on the one hand a centralization of highly informed decision-making and on the other an intense diversification of both products and markets.'[8]

Castells' sociological analysis details the growth and depen-
dency of global firms on information, describing the situation
as one of 'global informational capitalism'.[9] Essentially IT is a
management tool that enables corporations to grow and
compete in a competitive world. The so-called information
revolution, it is argued, does not lie in technical innovation, but
in the 'matter of differential (and unequal) access to, and control
over, information resources'.[10]

I have sought to locate IT within technical development based
on digitisation and within scientific management based on
efficiency and control and surveillance of production and consumption. We now turn to another technical development that
is integral to global informational capitalism – connectivity.

Internet – the vital connection

The Internet is, arguably perhaps, an astonishing achievement –
to put at least 60 million people (over 1% of the world's
population) in direct contact with global information and communication in the space of a decade.[11] Where did it start? In
national pride: Sputnik has been recognised as 'the father of
Internet'.[12] In 1957 Russia was first to launch a satellite to orbit
the earth; in response, the US government wanted to establish
a US lead in science and technology for military uses and
formed the Advanced Research Projects Agency (DARPA) in
the Department of Defense. In 1969 a communications network
was formed to link together research in several US universities
by means of telephone lines. This was ARPANET and was a
forerunner of the Internet.

Created by a researcher in the RAND corporation, the
Internet Protocol was designed to take any message and cut it
into little bits, each with its own envelope addressed to the
endpoint. Each little bit would then find its own way through
a telecommunications network and be joined back together at
the endpoint to recreate the message. The US military had

Table 1: The development of the Internet
After Everard[1]

Year	The development of the Internet	Events
1951		First commercial computer for sale.
1957	US Department of Defense establishes ARPA to bring together expertise in advanced computing in order to build more accurate and powerful weapons	Russia launches first satellite – Sputnik.
1962	Paul Baran's paper for RAND Corporation which explains how a system could continue to operate in the event of a 'small' nuclear exchange. He also advocated 'packets' and a system which could learn about the shortest routes through the network – called 'packet-switching'.[2]	
1965	Remote access to computers – in US, an ARPA study, independently in UK at National Physical Laboratory.	
1969	ARPANET established with a 4-site computer network, shortly to use the 1970 Network Control Protocol (NCP).	First manned landing on moon.
1970	UK's NPL network established.	
1971	France's prototype network Standards for remote terminal access (telnet) and file transfer protocol (ftp).	Telnet is used for Dungeons and Dragons, and later turns into Multi-user domains (MUDs) and MOOs (Mud Object-Orientated).
1972	First public demonstration of Arpanet. Vincent Cerf chairs Internetworking Working Group (INWG). First email (and probably use of @ for addressing). Introduction of dial-up services for remote terminals.	Lots of networks emerging on both sides of Atlantic. First commercially available 8-bit micro computer – the Intel 8008.
1974	TCP – a new protocol, the Transfer (or Transmission) Control Protocol, by Bob Kahn and Vincent Cerf. ARPANET begins to move from being a local area network into an inter-net of connected networks.	
1976	Information first able to be moved through all kinds of network (telephone, satellite, wireless, etc.)	
1980		British Library research project setting up and running electronic journals.[3]
1981	Computer Studies departments in US build CSNET.	

Year	The development of the Internet	Events
1983	TCP/IP used on ARPANET, NCP discontinued. ARPANET splits into MILNET and academic-only use, with a gateway to CSNET.	
1984	Japan network using UUCP.	First Apple Mac.
1987	National Science Foundation fund a 'backbone' for networking.	
1988	ARPANET began to be dismantled.	
1989	First relay between commercial email carrier and Internet (at Ohio).	News about China's crushing a democracy movement circulated via dissident groups on Internet.
1991	Access to file and information improved by introduction of Gopher (menus) and WAIS (search).	
1992	Number of hosts exceeds 1 million. World Wide Web launched, developed by Tim Berners-Lee at CERN. Growth goes from 0.1% of Internet traffic in March to 2.5% by December. First graphical interface with Web – Mosaic	Internet becomes publicly political with US Information Infrastructure Act.
1994	Commercial users outnumber academic (2:1). US Domain Name Server starts charging $50 for registration. First Internet access provided by commercial suppliers (instead of dial-up).	First Internet-only (using only TCP/IP) website on UK's Joint Academic network (JANET).
1995		At G7 in Brussels, Deputy President of South Africa, Thabo Mbeki, urges use of Internet to keep people informed of their own economic, political and cultural circumstances.[4]
1997		Asian countries see Internet as political when ASEAN seeks more of their material on it and use of their own languages.
1998	Internet users number more than 50 million (some say 100 million). End of US Department of Defense involvement with domain names (DNS).	

1. J. Everard, *Virtual States: The Internet and the boundaries of the nation-state* (London: Routledge, 2000), pp. 11–23.
2. P. Baran, 'On distributed communications networks'. *RAND AF* 49 (1962), 638–700. http://www.rand.org/publications/RM/RM3420
3. Author worked as Principal Research Officer on this.
4. T. Berners-Lee, *Weaving the Web* (London: Orion Press, 1999), p. 110.

feared that terrorist activity or a bomb could destroy fixed routes and so wanted a system where the message would get through, even if parts of the network were removed. With the Internet Protocol, provided that the system 'knew' about the existence of an endpoint, computers could be added or taken away at any point on the network and new communications lines included or existing ones removed.

Even before the worldwide adoption of the Internet protocol, email was in widespread use among academics, using a variety of public standards and proprietary software but requiring translation at key entry points to networks. Tim Berners-Lee invented World Wide Web (WWW, or simply 'the Web') in order to provide rapid international access to reports on the physics experiments done in the international research centre CERN. The Web comprises 'pages' which consist of information and also links that can take the user directly to other 'pages'. Its popularity grew when Marc Andreeson, then at the University of Illinois, created Mosaic, a graphical interface that used pointing and clicking, and which also made it easy to incorporate graphics and sound. The Web actually required the Internet and so its popularity led to the worldwide adoption of the Internet Protocol for linking computers around the world.[13] The combination of the Internet with World Wide Web meant that anybody could add a computer onto the network and then link to and from information on that computer in an easy point-and-click manner. In that sense, anybody could now be a 'producer' – and this was a major factor in the Internet's rapid growth.

Throughout this period of Internet development, businesses were connecting themselves up internally – as indicated in the earlier discussion on the growth of global corporations – and with their distributors and suppliers. Gradually the whole supply chain was computerised so that waste was reduced and new products were introduced and distributed more quickly. Retail outlets and their suppliers began using barcodes in order to record every item sold. This meant the retailers could order enough to avoid running out of stock on the shelves the next day, but not so much that they had to store unnecessarily, thus avoiding overinvestment and waste. (This is called 'just-in-time management' and it aims to ensure the product becomes available to customers just at the time they want it.) This systematic development of an internal information system together with the links forged along the supply chain only needed one thing to turn the whole of the computerised system into a consumer service – a link between the distributor or retailer and their customer. The Internet provided that link.

Virtualisation – transcending the here and now

As a fourth lineage, I want to present a more abstract aspect, but one that is a strong theme in much writing about information technology and Internet, and in science fiction. This is the idea that technologies have been developed to overcome the particularity of persons tied to space, time and body and to make aspects of a person available to others. Writing, for example, records events and ideas both for the writers themselves (a 'virtualisation of memory'[14]) and for others who are not present to the writer at the time. Many of the purposes of introducing information and communication technologies fit into this scheme.

Some look at the entire history of technology as a continuous striving to escape the 'here and now' – from the development of tools for exchanging know-how to the arrival of the knowledge society, and from the exchange of goods to the symbolic exchange of paper and digital money and e-business.[15] There are many ways that technology might be seen as an attempt to transcend 'here and now', and the two I consider here are the technologies that support the separation of information and knowledge from an individual and the technologies that support tele-presence or multi-presence. The former enable information and knowledge to be elicited from a person and made more widely available. The latter enable the presence of a person to be more widely available.

The process of separating information from a person has been an aspect of human existence from earliest history: to leave signs for others, to record events or to carry out some form of administration – whether notched in stick, painted on cave wall, carved on stone, scratched in wax tablets or scribed on papyri. In this sense, there has long been no necessity for an oral memory. With the introduction of writing for a whole class of readers/writers, rather than just by a group of scribes or priests, around 400 BC, a greater range of thought could be recorded and passed on. Later, the invention of printing greatly facilitated dissemination. The introduction of digitised analysis then extended the usefulness of stored information in that it could be searched and systematised much more easily. Issues of ownership and the right to use information existed before, but the possibilities opened up by digitisation raised all sorts of new questions. In the 1980s, for example, there were widespread debates in client-based businesses as to whether the staff's cultivated relationships with clients belonged to the employees and partners or to the company. If the latter, then any information proceeding from them could be put into a corporate database and used. After much resistance among some, the general rule was established that they belonged to the company, and staff often found their personal contacts targeted for business in other ways. Information that had been passed to an individual person was now made available to others. The continuation of this trend is the call for the knowledge possessed by individuals in a corporation to be stored in systems and made available to others rather than having it walk out with the individual or disappear on their death. This knowledge could be explicitly articulated by individuals and so stored, or implicitly derived from their actions; for example one might store the digitised movements of a particularly graceful dancer, and then train others to move in that way.

The other line follows the development of means to enable individuals to become present over a greater space and time than previously. These means range from simply increasing speeds of transport (allowing a person to be present physically in more places in less time than hitherto) through the provision of partial presence via telephony and tele-conferencing to the development of immersive virtual reality that allow a person to 'act at a distance'. One such example is tele-surgery, where the surgeon is in one location but the operation is conducted, via the Internet, on a patient in another location. The techno-logical transcendence of time and space may take various forms – what is transmitted to another place may be the voice, a visual representation of a person or written communication – but in all its forms, the goal of tele-presence is for someone to respond in real time to what is going on at a place that is not 'here'. This line moves towards ubiquitous presence – being present in several places at the same time.

Is cyberspace a new reality?

How has the metaphor of space arrived into the information held on computers in nodes on the network? In 1970s, interfaces moved from textual commands to a picture of a desktop with files, folders and ways of accessing the files, first made widely available on the Apple Macintosh in 1984. By use of the small piece of additional equipment called a 'mouse', one could point at an object on the screen, select it, appear to hold it and then move it across the screen – 'drag-and-drop'. More significantly, windows appeared to open onto views of a world that pre-existed – or things were brought into view by opening them into the window. By these 3-D metaphors, we construct a spatial model of our computer world in our heads.[16]

In webworld, geography is relatively unimportant. In fact one could almost define cyberspace as when the physical location of information, services, programs and people are unimportant – if it becomes important to know, you are not in cyberspace. Even if you were to know, the digital object or database could easily be moved to another location or stored in memory on a satellite – undetected and with no consequence for the user. There has to be some storage medium, the data must be physically located in some place, but where does not really matter. When first using the Web you might look at the addresses and see that the page of information on the screen comes from the USA, the next from a non-profit company in India, the one after from a library in Australia and – soon it matters little.

When you consider that physical location is unimportant, the metaphor of space is even more striking. Is it legitimate to describe an untidy accumulation of databanks, libraries, odd files of information and stuff (much of which we would probably have thrown away by now) as spatial? At its inception, the Internet used only text and was a very hierarchical structure. The introduction of the World Wide Web, with its hidden links between documents, recreated the strong sense of movement within another space, in contrast to the idea of moving between physical computers with specific site names. Tim Berners-Lee, inventor of the Web, constantly emphasised the word 'space' in his first proposal for funding in 1989.[17] In particular, he wanted everyone to be able to produce and share material easily, putting it into a shared 'global data space' with links. This might be one of the origins of the concept of 'cyberspace' but the word itself comes from a novel of futuristic science fiction called *Neuromancer*, by William Gibson, and refers explicitly to a data space.

When we use the Internet in its present form, the spatial metaphor in our minds is nurtured by the visual design cues as well as by the structural shape of the information, which creates its own logic and geography. Websites have addresses, welcome entrances and maps for locating the content; search engines help you to find what is where. As you 'move' around there is a real sense of visiting different locations and this creates the illusion of another space, a different reality.

Having existence in more than one space is not a new idea, and multi-levelled reality is part of human experience, Wertheim argues.[18] In particular she points to the dualistic theatre of reality in the Middle Ages: a physical space described by science and a spiritual space.[19] The situation is, she claims, quite similar to the present dualism in space: the material world described by science and the new emerging world called cyberspace. Between them there are connections and resonances, particularly through equipment and people, but they operate as two parallel worlds.[20]

Those spaces – material space, spiritual space and cyberspace, are substantial metaphors and easily identified, but also we use this term metaphorically in many other areas of life. Space is invoked in descriptions of relationships (personal space, room to move, etc.), of mental processes (space to think), and in science (evolutionary space, phase space), to give just a few examples.[21] These spatial images derive from two factors: contingency – things could be otherwise – and identification of features that we can map onto spatial dimensions. This mapping can be done visibly, as in 3-D data visualisation tools on computers, and imaginatively, as when in our relationships we map independence, locus of control and decision-making into a virtual space.[22]

Cyberspace, however, is more like a travel-and-location space than these scientific or psychological metaphors in that it uses a click and a link to move from one place to another, albeit the travel 'takes no time'. Its dualism is emphasised by the strong divide between what is accessible through Internet and what is not. As patterns of access change and cyberspace becomes the first port of call for more and more things, other routes are less well used and less well supported and so become more distant than before. A simple example is students' use of library resources; if books and papers are not indexed (and, increasingly, available) electronically, students are much less likely to find, use and refer to them. With lower usage (and government pressure for the digital), support reduces and slowly the paper-based systems drop out of use along with what they index. Early IT and Internet stuff is not available electronically (with some notable archiving exceptions) – which has led to several histories beginning much later at the point where digital records started. There is a strong divide between information and services available in cyberspace and those not. This emphasises the separateness of the two spaces – but the material world keeps intruding on the cyber one in a number of ways.

The stronger the connection of cyberspace with the material world, the more space and distance matter. It might be re-volutionary to say that 'time is timeless' and talk of the 'death of distance', but in fact the whole of cyberspace is supported by a material world that imposes itself undeniably. The dualism evidently breaks down when we look more closely at use of the Internet. Here there are connections with the material world through the physical storage and equipment with which we interact with cyberspace, in the invisible activity of others as they slow the system down by their activities, and in the signs of other selves I see in my interface.

The physical aspects of cyberspace end with the equipment that I use and, although messages may find different routes through Internet, I am still dependent on this equipment and a few key links that cut me off when they do not work. Whether the problem is caused by malfunction, or service provider or virus, I become very aware of the material world at this point. The speed that I can get access to sites in the USA alters radically when 'America wakes up and goes online'. At this point it is useful to know if a site is in the USA so that I can return at a more amenable hour instead of abandoning my attempt to access data as a lost resource. Our expectations of speed and frequency of contact are mediated through a sense of geography. My work colleague calls me up on the phone and says, 'I sent you an email ten minutes ago and you *still* have not responded.' Most unknown contacts from outside work, however, would be quite happy with a 24-hour response.[23] Distance and time are still alive when one is working in cyberspace. Nevertheless, I will argue later that the engage-ment and absorption of our interest, the quantity of content and possibility of contact, present a new space with new habits and norms of contact and communication. To give one small example, when researching for this book, I noted some interesting research and typed the name of the persons who had done it into a search engine. Following the links, Web gave me their email addresses, telephone numbers and addresses. Within a day I had spoken to all of them on the phone in their working hours, having made initial contact by email. An hour after that, several had sent papers to me and I had them printed out. I felt that I was operating in a new space–time compared with hitherto, with a window of my computer into it and the phone and printer as adjunct. As Levy has argued, all new technologies modify space–time in that they change what is thought of as 'nearby'.[24] In conclusion, I do use the term 'cyberspace' in this book, but stripped of any utopian sense, to describe the modification of space–time by use of the technologies for information and communication that are on Internet.

Locating the ethical debate

Many ethical discussions have centred around the use of the word 'cyberspace'. Often they take in aspects of digitisation, information control, Internet, multi-presence – and maybe several more, and this has tended to result in confusion of issues. Different lineages have developed their own sets of practices and norms: of what is acceptable and what is not. When lineages cross, conflicts of practice emerge, such as when the academic world of technical development comes up against the commercial world of control and the question is raised about what line to take with programmers who hack into software to demonstrate competence and prowess. Access to digitised information crosses different established practices. The way societies handle pornography, for example, ranges from exclusion at boundaries, restriction to certain physical localities or access restriction by age or time of availability – but the Internet circumvents these patterns. The lineages can therefore help identify where ethical conflict and debates will occur.

More important perhaps, is the recognition that different lineages lead to different utopian visions, whether explicitly announced or implicitly assumed. To draw some quick caricatures: digitisation points to an analytic world of data with an overarching aim of understanding how things work; scientific management's objective is complete control over uncertainties and efficiencies, so resulting in greater economic prosperity; the Internet in its communication aspects conjures up warm cosy visions of a harmonious world where social and political differences are overcome; and the transcendence of the 'here and now' projects a vision of a world where people can escape the finitudes of the body and – ultimately – of death.

Other voices, however, reflect that even if these goals were desirable, the utopias do not necessarily follow: just because we have the data to understand something does not mean we will act on it in the right way; better communication does not automatically lead to harmony (one might be expressing antipathy, hatred and exclusiveness); greater economic prosperity might be for some and not for all; and as people are dispersed out of themselves, there are personal costs – it has been suggested, for example, that people experiencing the effects of such dispersion feel the need to re-concentrate themselves into the 'here and now', and that participation in 'dangerous sports', such as skiing, windsurfing, bungee jumping, parachute-jumping, may be an attempt to do just that.[25]

An awareness of these threads of the story helps us to appreciate the deeper forces and motivations as well as the possible reactions to them. Tracing the lineages, however, is only groundwork. We need something more to guide us into examining ethics, and that is the space and context in which they occur; the *social spaces* in which they act.

From the discussion of historical lines and cyberspace we can see that information technology involves governments as supporters and users, corporations and organisations as investors and users, and individuals, whether as workers, consumers or citizens, also as users or non-users. One approach to ethical discussion has been to separate these levels and consider the ethics appropriate to the size of organisational unit (mega-, mesa- and micro-).[26] For example, the issues raised by extending

the use of scientific management in organisations and government are likely to be different from those affecting whether or
not an individual can afford to pay for the services offered.

Information technology is about more than information: in
Chapter 1 we saw it covers four main areas of use – information,
communication, virtual reality and cyberbusiness (or e-
business). Each of these can be considered in terms of the
organisational units – which gives us give a grid of possible
issues to explore. We could look, for example, at the impact of
virtual reality on an individual or the use of communication by
international crime organisations.[27] Or we could consider how
all sizes of organisational units are concerned from different
perspectives on, say, the disclosure of personal communications: government from the view of prevention of crime,
organisations for their protection of use of time and release of
information, and individuals for what they can expect to be
'private'.

Furthermore we could then add into the grid the kinds of
issues that could be raised (economic, security, environmental,
etc.). We would then have a 3-D grid, with the uses of cyberspace interacting with organisational unit size interacting with
any number of issues. The access question, for example, may
emerge for governments at an economic level, as they determine
whether or not others abide by rules for trade (for example the
Berne Convention, to prevent abuse of copying without paying
copyright fees); or, at organisation level, to guarantee sufficient
access so that all who want to go onto Internet can do so
without delays; or, at an individual level, to ensure that no one
is excluded from participation and hence from public facilities.

From these examples, the number of interacting factors can
be seen to be vast. To take every issue, unit size, use of cyberspace, and then to work through systematically is far beyond
the scope of this book – or indeed any individual work. Parts
of the grid of interrelated factors are being explored and the
reader can follow these up from the annotated bibliography at
the end of the book. Here we will concentrate on the social
aspects of individuals, their communications and relations –

an approach that does stress the uses of information and communication, but does not ignore cyberbusiness and virtual reality.

Talking about cyberspace inevitably brings in technology, and so before we turn to the social spaces, the next chapter investigates the approach to technology in general which I want to use in this book.

chapter three
Technology

Technology and social change go hand in hand. How do they relate to each other? We examine technology's characteristics, and make some points about it:
- technology is not just a tool, nor just an alien force imposing novelties on society;
- it is neither good nor bad, nor is it neutral;
- it is 'malleable', i.e. open to adaptation, not predetermined;
- it brings unforeseeable change;
- people need to be able to change its direction when necessary for the common good;
- it needs the use of 'invitational' rather than 'limiting' power; love of neighbour calls for a participative approach to technology, especially in the interests of those known to be adversely affected.

SOCIETY CANNOT BE understood without its technological tools.[1] Historical periods are strongly characterised by their tool-making capability and some – such as the Stone Age and the Iron Age – are named in these terms. The very close relationship between technological and social change is evident throughout history and perhaps most conspicuously in the Industrial Revolution. The same relationship exists at the level of specific technologies such as the motor car, the electricity grid, the telephone and the computer. It would be hard to imagine Western culture without its road, electrical and telecommunications infrastructures. But which comes first? A new technology or a societal call for it?

Neither tool nor alien force

A commonly held view is that we use technology merely as a tool; that society develops the technology it needs and then uses it to produce goods and services for the creation of wealth and for human culture to flourish. In this picture, needs and wishes come first and the technology simply fulfils them. But this view completely ignores scientific exploration and inventiveness. People do not always know what they want because, from the standpoint of what is currently available, they cannot always imagine what might be possible: scientific endeavour leads to technological innovation and holds out new ideas for consideration. However, this can appear to be taking place as an activity separated from the rest of society – and so it feels as if new technologies are constantly being imposed on us from somewhere outside our culture, as from an autonomous alien ruler (so-called 'technological determinism'). Many people feel that genetic manipulation and cloning have arrived, uninvited, in this way. Much of the rhetoric of information technology adds fuel to the fire, presenting technological innovation as an alien saving force to which we must adapt or be left behind.[2] So, if we use technology as a tool, then it is available to us when and if we decide to use it; if as a force impacting on society, then we often feel that we have no choice but to modify our patterns of living to fit.[3] How do we resolve this sense of simultaneous choice and no choice; of being in control and also out of control? This requires recognition of the various factors that influence the development and adoption (or rejection) of technologies and the attitudes towards them.

First, there does exist a strong interaction between the demand for new technological tools and their application, but it is not a simple, linear, causal relationship. Of the many technologies that are invented and marketed, there are some that the general public don't choose to adopt or use widely (for example Prestel, a TV teletext service in the UK, or natural energy sources for power). On the other hand, people want inventions that are difficult to provide (such as low-energy, non-polluting cars or software that does not contain bugs). Whether as individuals or organisations, people feel they get some degree of choice, but certainly not everything they want or need.

Second, technology usually gets invented and developed in organisations within which some ideas are easier to explore than others. For example, the introduction of the World Wide Web was bedevilled by organisations interested in developing Internet services but more concerned with centralising ideas of databases and notions of their own polar position – and indeed much of the inventor's vision of a system with distributed capabilities has yet to find a way of being implemented.[4] Conversely, the requests that organisations make of technology are formulated from within a set of social relations – the economic one of capitalism being strongest – and 'big' technologies suit big organisations better than distributed small ones. Technology depends critically upon its relationships to the institutions employing it.[5]

Third, technologies may take on different aspects according to where they are used. At home they may feel like tools that are chosen and enjoyed; the same technologies at work (depending on how they're used) may feel compulsory or even oppressive.[6]

Fourth, the development of infrastructures, such as the road network and electricity grid, feel big and beyond individual influence: they change life around us, irrespective of people's personal choices. Many technologies used in a consumer society are large-scale systems with small, purchasable tools that plug into them – the Internet is a good example. People buy the equipment and resources to get onto the Internet, but equally, probably feel it would carry on without them if they did not do so.

Technology, therefore, cannot be viewed simply as tools to

be called for and used, neither is it an alien force; in reality there are a set of complex relations between the sponsorship of technologies by commerce, government and the military and the specification and purchase of technologies by organisations and individual consumers. Technology does not come as a finished package, but is developed and shaped in the context of needs and inventiveness within societies. The corollary is that the actions of individuals and organisations are not without effect on technological development. In information technology and cyberspace, technologies inhabit a social world where they are engaged in a two-way process of adaptation and accommodation – a process wrought by social interaction rather than between a single person and the encountered surface of a technology.

Neither good, nor bad, nor neutral

Technology is purchased because it will give benefits to someone or some organisation. It therefore looks good to those introducing the new tool, but may not be seen that way by everyone involved. It was in the 1980s that a bank first gave computers to all its staff so that it could monitor and optimise every deal and transaction – further down this line, it now has accounts and forms on the Web. Before the 1980s, the bank manager made the decisions: he (and it almost certainly was 'he') had autonomy to do so. The arrival on his desk of a computer that told him whether or not to make a loan brought a considerable transformation in his job – one he might not think was good. The more recent move to Internet banking has led to local bank branches being closed and the laying off of staff – who may not see the change positively. The same tool has different values placed upon it. The same technology can be seen simultaneously as both 'good' and 'bad'. The example also highlights the effects on social relations – both directly, among those within the organisation and with those outside it, and indirectly, as a result of new dependencies on programmers, equipments, manufacturers and other monetary exchange.

Technology changes social relations, but not only through use. If I have a gun in my hand it changes things, whether or not I use it. Even the possibility that I could and might use it alters the social situation. It is not simply a tool, like a pencil-sharpener. This idea is clearly apparent in the existence of weapons that are not intended for use but pose a powerful threat – such as nuclear armoury and biological Trojan horses. These unbalancing effects of technology on social relations result in what Kranzberg called his 'first law': technology is neither good, nor bad, nor neutral.[7]

The imagined neutrality of technology emanates from a narrow focus on the purely technical and functional aspects of technological devices – a hammer for hitting nails, a car for transportation, a phone for communicating. As a new technological device gathers a whole industry around itself, its function becomes established and immutable, and that fosters the concept of neutrality – even when the device has altered the character of the social landscape. Of course it suits the industry to insist on neutrality, because this avoids the need to respond to challenge, for example over safety or environmental standards – a matter to which we will return. Much of information technology, though, as yet has no such established functions, and so I believe that herein lie many opportunities for non-specialist participation in the development of the life-world where all may flourish.

Malleable, not predetermined

Even when technologies are considered in terms of a body of knowledge about how to do something (how to make fire, how to build a means of crossing a river, etc.) or as an integrated system such as the utilities of electricity, gas and water,[8] we still think of them in terms of 'function'.

These functions are not predetermined – the society develops them. The home computer, for example, has major functions as a communicating device – for getting information, for sending and receiving messages and for preparing text for printed communications. But in its early days, many imagined that its function would be controlling household appliances to reduce cost and maximise efficiency, much as in manufacturing industries. We can deduce from this that people have been involved in directing the development of information technology. This means that what people think and do is by no means unrelated to the shape of the technology they get.

One thing is for certain – the function of the Internet has already changed: originally a data network, it was literally reinvented to become a communications one. Had it been obvious that communication and information would be integrated on Internet, few would doubt that Bill Gates and the whole of his company, Microsoft, would have spotted it coming. He didn't, and had to turn the whole company around to meet the changing environment of his business. Both the French Minitel and the Internet were originally designed for the distribution of data, not for communicating. When, in the early 1980s, the national telephone company in France, with financial assistance of the French government, distributed millions of terminals free of cost, the move was intended to provide access to a centralised data service. The users, however, demanded communication and soon the main use was online chatting for fun (and sex). In the UK, the General Post Office (at that time comprising both national telephone and mail services) sought legal action against firms that they discovered passing messages via data networks instead of using the GPO's phone or mail service – thus breaking their contract.[9] At the time that Internet standards were being developed, many computer companies laid off the teams that were working on communicating PCs, not believing that there was any future in them.[10] Yet today it is difficult to think of a computer without a communications facility (mine has all the software for telephone, answerphone and fax as well as Web, email and file download capabilities). The point is that 'functions' are created along with the technology and shape the development of both the actual device and the lifestyles it makes possible. Technology is far more malleable than it first appears, as these examples show. It also illustrates that the future is not quite as predictable as one might expect.

Looking at established technologies only in terms of their function has a particular drawback: other aspects such as appearance, ease of use, ease of repair and impact on the environment become secondary. But this is a peculiarly North Atlantic view, not necessarily found in all societies; for example, the religious and the aesthetic play an equally important part for some Islamic countries and this can appear very confusing to those who think primarily in terms of 'function'. There are, however, instances in which these other aspects play a greater part, even in the Western world. In the home, multiple considerations surround all the apparatus of living: it is difficult to see it purely as 'a machine for living' (to use Le Corbusier's expression). In some societies (certainly in the UK) the car does have other qualities that compete with the primary, functional one of transportation – in particular, it is a signifier of status or position. Any technical device has a set of roles that include not only its function but its social impact and the lifestyles it facilitates. One of the ways in which a technology can be shaped, changed or developed is by broadening the view to include the full set of roles: to see the other roles as intrinsic rather than peripheral.

Unpredicting the future

Technology changes things – if it didn't, it wouldn't be of any use! Exactly what it changes is not always what people expect or desire. No one quite knows in advance how technology will be picked up and used, particularly in the consumer market. A major predictive survey of 1895 completely failed to anticipate the car and its impact. The future of information technology was similarly unanticipated, even by the experts: 'I think there is a world market for maybe five computers' – Chairman of IBM, 1943; 'There is no reason for any individual to have a computer in their home' – president of DEC, 1949; 'Computers in the future may ... only weigh 1 ton' – *Popular Mechanics*, 1949.[11] When people pick up and use technology, they change their patterns of living – they do things they couldn't do before and abandon other things that are no longer necessary. The cumulative effect of this is very hard to predict; in fact for most technologies it is nearly impossible and this is what leads to surprising results.[12] The use of chlorofluorocarbons (CFCs) was deemed so much safer than the fire-prone expellants of early aerosols that no one asked, 'What will be the effect on the atmosphere when every person in the country uses two or three a year?', and it almost seems unreasonable to have expected them to do so. People launch things without fully knowing all possible ramifications. When the Internet began, the banks probably did not imagine that a data network might result in a wholesale change in their practices and offers to customers, yet that is exactly what is now happening. Even if present knowledge of all the affected areas was perfect, there would always be too many interconnecting factors for all the outcomes to be predicted. The result is that the introduction of new technologies unpredicts the future – we simply don't know what will happen.

Another 'unpredicting' effect results from the almost universal goal of organisations to optimise efficiency. This they do to such an extent that if anything goes wrong, their service rapidly destabilises. This can be seen in tightly coupled systems, where the drive for efficiency has instated computer systems that handle complex calculations to optimise everything that is happening. Such a system is found with air travel in the USA,[13] where the door of an aeroplane closes, yet the airline knows that a passenger is running from a connecting flight and is only three gates away. Whether or not to keep that door open was a real-time decision taken by a computer and based on complex modelling. The computer knew how many people were late for each flight and by how many minutes. It would have considered the distance to the gate, all the times of the next available flights to the same city, and the likelihood of new delays at the far end. It would have calculated the crew's shift times (two or three minutes late in starting and they might run over legal limits, so that whole plane has to be delayed with a crew change); it would know about all the different weather conditions across the routes and how they might slow and reroute traffic; it would be making calculations to ensure that there would always be a safe landing spot within reach (calculating fuel, flight requirements on one engine etc.). The result is an efficiency that means that fewer than 2 per cent of the planes are idle at any time. But when there's a problem, there is no slack to accommodate shocks or failures, and small events can quickly run out of control. Tightly coupled systems quickly destabilise, because of, rather than in spite of, the attempt to control every aspect.

Technological 'repentance'

If technology unpredicts the future and it turns out to be a future that is not wanted, then we need processes for changing direction. The unpredictability of technological innovation and implementation is a problem, particularly for those in the Judaeo-Christian tradition who believe that God made humans creative but not omniscient. If God made human beings to be able to think and introduce new ideas, but not able to see the consequences of so doing, then how can they be 'responsible'? It can seem as if 'God' does not want innovation to succeed! This was vividly illustrated on one occasion when I was taking part in a TV show on technology and the environment. A man in the audience claimed, with a voice of outrage, that God was awful and vengeful and took delight in making everything go wrong just when people were trying their best. He referred particularly to the way in which CFCs (chlorofluorocarbons) had to be replaced in aerosols because they were damaging the ozone layer, even though they were originally introduced as a

life-saving alternative to the much more flammable prede-
cessor.[14] As I attempted to explain, it's not that God frustrates
all we attempt to do, but that 'our lives are controlled by
decisions we made when we did not know what we were
doing'.[15] This is true at every level, from everyday personal
interactions (in which sharing can be an 'imprudent
exposure'[16]), through life decisions (such as the person we
choose to marry and which job we take and city we live in) to
designing new technical equipment. We don't know the outcomes of our decisions. Of course it would be possible to do
nothing at all and so take no risk, and indeed this is often
argued for with regard to technology (mostly in reaction to a
new idea). If new ventures are risked, then they should certainly
proceed only with the utmost knowledge that can be mustered.
But perhaps more importantly, there also needs to be a way of
pulling back or changing direction if it turns out to be detri-
mental to the well-being of human beings or the environment
on which all depend. Because this is a corporate activity, it
follows that there is a requirement for ways of talking about
the problems and then introducing processes for change. This
will not be easy. Politics is governed in the USA, for example,
by 'scandal', proceeding by oppositional and negative undermining of credibility.[17] This is not the environment in which
changes of direction in technology can be made – it is too
threatening for those in politics. We need ways of making technological 'repentance' real and acceptable, so that it should be
considered normal to reconsider where the technology is
headed and, if necessary, change direction. This is what I mean
by technological 'repentance'.

There have been instances of such 'repentance', and they
include the introduction of speed limits in order to reduce
human death rates; the redesign of boilers to reduce human
death on riverboats; the introduction of lead-free petrol to
prevent exhaust fumes causing brain damage in young children;
and now the discussion on carbon dioxide emissions causing
the 'greenhouse effect'. These have all been contested because
change is not in the interests of those benefiting from tech-

nologies with established functions (and so revenues), for reasons of 'efficiency' or direct cost. Any purported change involves players in relations of power – and this leads us into one approach to exploring technology, its introduction, implementation and change – not least because every religion expresses views on the use of power.[18]

Relations of power

Technology as good, or bad, or not neutral, can be viewed by different people as promise, or threat, or relations of power.[19] It looks like promise to those it benefits – particularly in business services and a few industrial sectors (wholesale and retail trade, finance, insurance and real estate) who have invested heavily.[20] These are primarily organisations who need to know exactly what their customers are buying. It looks like a threat to those whose jobs have been removed (mainly blue-collar workers who have been replaced with automated machinery and robotics), and to those who do not believe in the total innocence of all those in charge of surveillance – whether they are following our movements by analysis of mobile phone use, credit card purchases or number plate. The promise and the threat are often set against each other: utopian prosperity versus the neo-Luddite scenario where benefits are for business and not individuals. The promise is brought by technology; the threat also. So in both cases, technology usually frames and limits the debate. An alternative path is to view technology in terms of its relations of power. I find this more helpful because that takes us back to relationships in society, between and among organisations and consumers, governments and their citizens. I also prefer it because the right use of power is a theme that all religions take seriously.

Power is usually defined as limiting choice of action for others. 'Power is the ability to reduce, limit or eliminate alternatives for the social action of one person or group by another person or group.'[21] This is one definition offered by the sociologist Dahrendorf.

Relations of power are abundantly evident in the realm of information technology. British Airways was taken to court and found guilty for ensuring that its holidays and flights came up first on the screens of the online services in travel agents. This was considered to give them an unfair advantage because people generally don't scroll down to the other options. Microsoft was taken to court and had to retract their bundling of Internet Explorer with the Windows operating system, which had meant that purchasers of personal computers had no need to obtain other Web browsers such as Netscape. These are just two examples among commercial organisations. Relations of power are also embedded in the design and building of technological systems. Some believe that selling software to consumers that works poorly and 'crashes' frequently, without any offer of recourse, amounts to the use of power for profit – particularly when the consumer has no way of adequately checking what they are buying.[22] Even if the software performs well, some sociologists would also consider that the inclusion or exclusion of functions and facilities within it, constitutes use of power.

> It is in this sense (of power) that computer programmers, the designers of computer equipment and developers of computer languages possess power. To the extent the decisions made by each of these participants in the design process serve to reduce, limit or totally eliminate action alternatives, they are applying force and wielding power in the precise sociological meaning of these terms.[23]

Some readings of the Ten Commandments point towards the non-use of this kind of limiting power. The Israelites in Egypt had only experienced the abuse of power, limiting what they could do and not do. One aspect of the Ten Commandments was their aim of creating a new society with new relations of power – whether power expressed in physical strength and health, handsomeness/beauty or economic wealth: 'thou shalt not . . .' take advantage of the weaker among you, whether because of age, or being a foreigner, or because you can get away with their spouses, cattle and other goods. New relations of power were to be the mark of 'a priestly kingdom and a holy nation'.[24] They extended, for example, into ensuring that enough was left on fields after harvest for the poor to glean, rather than keeping it all to oneself. They extended to God's reluctance to endorse the appointment of a king. They extended to reconsidering the concepts of ownership (i.e. God's). A new identity meant new patterns of relating to one another. The prescriptive law was the start, in response to the fact that their minds were shaped by the society of which they had been a part – as we can read from their desert grumblings and thoughts of return. The question was, how to transform a people? The objective was to create a new people relating to God and to each other. Jesus takes the Law and brings it back to its original purpose, picking up on the aspect of power, most clearly in the Beatitudes in his Sermon on the Mount.[25] Translating 'Blessed . . .' as 'You are in the right place if . . .', the Beatitudes can be taken to mean that you are in the right place if you are not in a position of exercising power to limit the activities and future possibilities of others.[26] Within relations of power, Jesus seems to be promoting the idea that one side is better to be on than the other. He does more than that, pointing to a positive kind of power.

All we have discussed so far is a negative power, but there is another, different kind of power – that evinced by Jesus and used in healing and giving life, in increasing the possibilities of action, rather than limiting or eliminating them. This power is invitational – and must be offered as such since, with any introduction of force, it reverts to being negative power. Jesus has and offers *invitational power*. This is the basis of the Calvin Center for Christian Scholarship's definition of technology:

> Technology is to be done as a form of service to our fellow human beings and to natural creation. This means that we are to develop technology in such a way that the blessings, riches, and potential God has put in creation are allowed to

flower. . . . Second, our technological activity should reflect
love for God and neighbor by expanding, not constricting,
the opportunities for men and women to be the loving,
joyful beings God intends them to be. Our technological
activity should increase opportunities for us to freely
choose and act, thereby contribute to society.[27]

We can use these criteria and ask whether information technology and cyberspace tools do in fact open up opportunities
for others or whether they close them down by means of financial cost, education, how and where investment is made,
infrastructure or straightforward exclusion. Similarly, they can
be applied to the design of products and systems in considering,
for example, issues such as how the software bugs will be fixed
in order to make computer programs work better – should the
source code of computing be available, as the proponents of
'Open Source' argue, and, if not, how do proprietary manufacturers ensure that their products are improved and how do we
participate in that process?

Shaping the future of cyberspace

Information technology and cyberspace alter both the structure
and form of relations people have with each other. The second
of the two commands of Jesus (to 'Love the Lord our God
with all your heart, mind, soul and strength' and to 'love your
neighbour as yourself') must therefore find new expression.
Our relating to neighbours is changed, but so is our solicitude
for them, as expressed in care and concern and in earnestly
demanding ('soliciting') on their behalf.[28] As neighbours experience a changed social landscape because of new technology,
our solicitude is to act on their behalf to challenge the things
that disadvantage them. The ethical demand presented by other
people therefore includes concern for their circumstances when
they become adversely affected by information technology.
There are many areas where this is the case, for example the

problem experienced by older people who are unable to get to a bank because they are unable to drive, and local branches are being closed with the expectation that people will use the Internet or phone banking. I am arguing that 'love of neighbour as oneself' demands a participative approach to technology that seeks the welfare of at least those known to us to be adversely affected.

Active participation is only one approach that might be taken by a Christian (in distinction from a citizen's) response. One might choose to oppose the technology entirely as part of an artificial culture constituted against God's will – as do the Amish. At the other end of the spectrum, some argue that there is unity between the co-creative activity of human beings and God's purposes which means there is no difference between God's desires and our activities, flawed though they be. Certainly some technical developers see the communication aspects of Internet in this way.[29] H. Richard Niebuhr characterises the relation of Christ and culture in five different ways, of which these oppositional and unity positions are two. A third position is that the cultural (and so technological) project and the faith project are two completely separate endeavours. The position that I assume is Niebuhr's conversionist one, in which Christ enters into culture through the activities of those who seek to 'do his words'. As such, the active participation on behalf of others in shaping and developing technology seems to be placed here.[30] My view is that technology is malleable, that there are opportunities to participate in shaping it, and that this is one possible Christian response within the context of loving one's neighbour.

I have argued that information technology is formed in a complex interaction between technical innovation and societal need and uptake that results in a changed social landscape. IT presents itself as devices to fulfil particular functions, and that is a consequence of social development. However, the functions or aspects of IT and cyberspace are not yet fully worked out and there are many possibilities for shaping the future. Neither the social landscape, nor the devices, nor the consequences of long-term and widespread use are predictable. Some of the issues and areas to be contested are considered in the following three chapters, Chapters 4, 5 and 6.

The Christian ground for exploring the issues in IT is one which stems from the command of loving one's neighbours as oneself. This demands not only a direct ethical response to their need, but also soliciting change in technology that impacts upon them. In particular, what is sought in this active participation is invitational power for the other: to increase the opportunities and choices for them to act and so be able to flourish should they choose to do so. In this position, controlling power has to be laid aside. The demand for change results in the need for new processes, which are discussed more explicitly in Chapter 7.

This is not a position from which to protect the status quo but one for choosing to use, not to use, to challenge, to oppose, to give meaning and integrate technology into lives, in a non-controlling way that supports invitational power and use in co-dependence with others.[31] This is not to be silent but to engage with cyberspace in such a way as to help transform it, for it is possible to shape the life-world that results from the introduction of IT. This is how I propose we engage with and shape cyberspace.

chapter four

Information and Knowledge

How do people interact with information? What is the relation of information to knowledge? What are the ethical issues that this aspect of IT throws up? We look at:

- the passage from data to information;
- the transmission of information;
- the changing value placed upon the content of information;
- the personal and social implications of competing sources and demands for attention.

THIS IS THE FIRST of the four chapters that explore the ways in which information technology is changing the social spaces, looking in turn at people's interaction with *information*, with *others*, with *self* and with *political participation*. We begin with data, information and knowledge and, having defined and distinguished the three, the aim is to identify the issues that relate to ethical questions.

First, we examine data and information: both the collection and storage of data, and its processing into information, carry many risks of error and raise questions about accuracy and truth.

Second, information technology at every stage of development has affected the transmission of information: we examine the concerns and the implications.

Third, information technologies are changing the value placed upon content, whether through the rights to copy and make money from it, or through the possibility of attaching direct monetary value to knowledge. There are some interesting questions surrounding the changing patterns of value of content, be it in form of information, knowledge, writing or music.[1]

Fourth, information technology is used to help grab and maintain our attention, both by its devices (phone, computer, etc.) and by its data (the 'personalisation' of mail, email, and soon digital TV). Competing sources and demands for attention carry implications for the well-being of individuals and organisations.

Informing process is error prone

That I am five feet eight and three-quarters of an inch tall is a piece of data. It becomes information in response to a question: will I fit into the cabin of this boat without bending down? Am I the person I claim to be when I present my passport to an immigration officer? Other data relate to my financial state and include records of payments and addresses where I have lived. The kinds of questions asked of this data may be about credit – do I have a suitable profile to qualify for borrowing money for a mortgage? Am I the kind of person who is likely to default on payments? The answers to those questions are information for the person asking.[2]

Concerns surround the accuracy of data and information and their subsequent use. Wrong data and information can change life chances for many people – as, for example, when it leads to a credit agency refusing a request for mortgage or loan. Incorrect information can be the result of wrong data, incorrect data entry or inaccuracies in data transmission and storage. There are also many points at which data can be in error. Incorrect data entry can lead to spectacular, headline-news mistakes, such as the one that affected British aerobics instructor Liz Seymour. She thought she might have drawn too much from her bank account when she returned from vacation but was surprised to see her bank statement claim she was £121 billion overdrawn. The bank notified her that she would be charged £2.5 billion per day in interest. When questioned, the bank said it was the result of a typing error.[3] Of course it is easy to see that this was an error and laugh it off, but it becomes much more difficult in cases where discrepancy is smaller and the size of the number is approximately correct – as happened when I was paid the wrong amount of salary because my employer's bank had mistyped the figure when it made the direct payment into my account.

Problems can also arise in the information processing of questions asked of a databank, in the coding of programs and in the subsequent transmission and use of the information. The programs that run any information processing are far harder to check than the data itself. A scientific example comes from the measurements of the thickness of the ozone layer over the Antarctic. The scientists had an error-checking program to examine data before it was placed into the databank. Because they assumed that the ozone layer would gradually thin, data showing zero thickness – i.e. the existence of a hole – was systematically thrown out as erroneous.[4] It took someone to ask a different question before the extent of the damage to the atmosphere was apparent.

Information programs are vulnerable, and not only to unforced error. As I was driving down a highway looking at the cameras that automatically check my speed and number plate, I was wondering: suppose I was on a contract programming the automatic sending out of the fines. How long would it be before anyone noticed that I had inserted a line that read 'if owner contains "David P", then do not send fine'? Or what if I had a boss who didn't like me and in vengeance I wrote him into getting a fine for very small infringements, while others were permitted the latitude agreed by the police authorities? And since I would, in all likelihood, have moved on to another company, who would be liable? We already know that programmers in banking performed an equivalent manoeuvre with the small remainders of money from percentage calculations and embezzled millions of dollars, passing them into foreign bank accounts. It took a long time to discover what had happened there.

As society comes to depend more on data to support the informing process, especially for decision-making, it needs to be sure that the data and the data-processing is accurate and free from systematic or random error. We are still a long way from that situation and there has been no evident attempt to institute in the data industry the levels of professionalism that exist in, say, the medical profession. In the UK, the Data Protection Act 1998 (DPA) addresses the collection of data, its transmission and the results of any processing, but not the quality of the programs that are used to extract and manipulate the data to be presented and used as information. Moreover, the data may be contested only by the individual concerned (the 'data subject') and wrong data can flow around the Internet in such a way that it is nearly impossible to locate all the places where it might be lodged.

Encompassed by the main thrust of the European Community legislation (of which the DPA is the UK implementation) is the need for data about individuals to be protected from access and circulation by those with no need for it. Under this legislation, only living people are protected: when a European citizen dies, their personal data is no longer protected. A different piece of legislation, the Computer Misuse Act 1990, seeks to make any unauthorised access to computer material and subsequent modification illegal. However, the original programming itself is not subject to the same degree of inspection – a situation that has prompted at least one initiative to insist that two programmers work on every piece of code.[5] In summary, people are largely protected in Europe, though not in the USA, from the transfer of data and information to those who might misuse it (access). However, they are not sufficiently protected from error throughout the system (accuracy) because the burden of proof is on the individual concerned and because such errors are not yet deemed to be so harmful that those responsible for them see the need to change what they are doing.

Information processing belongs in the brain?

Viewing the Internet as a massive data store causes misgivings among some, who believe people will lose the use of their memories and also, as a result, their creativity. In other words, they will cease to be (literally) in-formed by what they discover through asking questions. This is what the playwright Alan Ayckbourn reasoned: 'The e-mail generation feels they don't need to know very much because it's all on the computer. They don't need to learn about Edward II because [they] just do EdwardII.com. But when the mind no longer becomes the processor of information, you don't have any creative thought.'[6] He was perhaps aware of earlier remarks by Plato (c.428–c.348 BC) on the introduction of writing:

> [T]his invention will produce forgetfulness in the minds of those who learn to use it, because they will not practise their memory. Their trust in writing, produced by external characters which are no part of themselves, will discourage the use of their own memory within them. You have invented an elixir not of memory, but of reminding; and you offer your pupils the appearance of wisdom, not true wisdom, for they will read many things without instruction and will therefore seem to know many things, when they are, for the most part ignorant and hard to get along with, since they are not wise, but only appear wise.[7]

He went on to formulate his argument against writing (through the voice of Socrates) in a series of points, summarized here:

- education will suffer because it presents information rather than promoting thought;
- authorship will be difficult to authenticate;

- information security will be compromised;
- writing will be nothing more than shallow distraction, devoid of serious purpose;
- people will stop interacting with real people.[8]

The first point encapsulates Plato's main concern, which was for learning. The Internet is certainly changing the type of learning experience on offer, as did writing many years previously. Online courses are being introduced by universities throughout the world in order to offer students easy access and flexibility – so-called 'e-learning'. Early experience and research confirms what Plato believed: that there has to be sufficient opportunity for students to internalise – to gather the ideas into the memory – by means of discourse and debate. E-learning can work successfully, provided the student has enough interaction with their tutor and with other students in the group.

Plato's second point about writing – that it can be difficult to know by whom it was written and so the quality of the source – has been true of every innovation in information technology, from writing through printing to the Internet. Stories abound of people quoting from sources on the Web that turn out to be a school-student's essay, or taking extracts from an online discussion group as advice on whether or not to buy stock. The nature of the problem, however, is exactly the same as exists in paper and ink. School and college students have always been taught to examine the nature of the source and to consider the bias there might be in it: the same can be done for Web sources. Just as some sources are considered reliable in print, brands of reliability will emerge online.

Third, Plato identified the issues raised when material is taken out of its original context: these include questions over its veracity and whether the recipient is in a position to handle it. Such was the nature of the problem that arose over a radio interview with the spokesperson from the nuclear reprocessing plant, who said that of course they would get many leakages, as you do in all complex industrial plants. This may be true, but it caused much alarm because it was interpreted in quite a different way to the manner intended – which was to confirm that they knew how to handle the problems. There are, of course, numerous examples of types of information that would not be appropriate for everyone to have, including porno- graphic material, instructions for making bombs and subversive writing. Through the Internet, people do have access to content they did not before and this will raise issues, just as it did for the Church when the Bible was released into the vernacular.

Perhaps it is more difficult for us to imagine what Plato meant by his suggestion that writing would be a 'shallow dis- traction', but our perspective is one that takes in the evolution of writing into sophisticated forms of expression that include novels, scholarly works, poetry, etc. In Plato's time, in-depth thought and development of ideas was done person to person; the possibilities for using writing in this way had yet to be explored. Similarly, we may regard the trivial content of email and many Web sites as a 'shallow distraction', but the Internet is a young technology and we do not yet know what forms of expression will emerge or exactly how they will be used. However, it is also interesting to note that Plato's argument here is somewhat belied by his own work!

Leaving the final point for the moment (we will cover it in the next chapter), we can see that both Ayckbourn and Plato hint at something rather deeper than methods of handling the new medium and the successful acquisition of learning. Ayck- bourn fears that the new medium will cut out the information- processing that fuels creative thought. From Plato comes the idea that with a new form of information technology, people may apprehend information but not necessarily appropriate it: in other words, they will not acquire 'real knowledge'.

At this point I should say that my intention in the following discussion about data, information and knowledge is not to set out definitions of the areas discussed, but simply to explore what appears to be happening in them and how people are responding.

Knowledge is direct, information indirect

What is the difference between having information and having knowledge? One distinction is that information tells you about things that are not present – that are 'elsewhere and elsewhen', whereas knowledge is about things present to you.[9] Borgmann, following Bertrand Russell, compares knowledge by description and knowledge by acquaintance, that is, indirect and direct knowing. French and German, he notes, have separate words for these two ideas; indirect knowing (*savoir* and *wissen*) and direct knowing (*connaître* and *kennen*). Information is indirect, whereas knowledge is always direct. This is certainly the meaning that the Bible attaches to knowing – it is a direct experience, as indicated most strongly in the use of this word to describe sexual intercourse.

Direct knowing helps us to see things in a completely different way. How many times do we say, 'if I had to do that again I would do it completely differently'? When we started we had information; afterwards we had knowledge. The difference between the two was nicely illustrated by an encounter with an airline pilot. On a transatlantic journey, I asked to go into the cockpit and talk to the captain. While there, I asked him whether he ever overruled the advice given to him by the computer on route, altitude and flying speed. I knew it was possible to do so, but didn't know if they ever did. The captain smiled and told me he had done so on this very trip and then explained how this particular 747 had Rolls-Royce engines that were more powerful than the ones usually fitted. Normally this was irrelevant, but that day there were strong headwinds and he had calculated that he could fly higher and still make a faster time, despite the increase in wind speed at higher altitudes. In fact, he was going to save over half an hour for us on the journey. The astonished flight engineer then interrogated the captain about how he knew so much – since most pilots just flew the plane. The captain replied that he had studied aeronautical engineering at college and had always had an interest in getting the best performance out of his flying. Although the computer did not know about the engines on the plane, he did and was able to overrule its advice on this occasion. He had constantly used his knowledge in a wide variety of conditions, knew what was possible and safe, and could explain and justify its use articulately. Here was demonstrated a clear distinction between knowing about something and the direct knowledge that came from using it.

Many types of knowledge can be communicated to others – basic know-how, the best and quickest ways to do something, the top ten tips you learned and wish someone had told you when starting a new job. As discussed within the 'out of body' lineage in Chapter 2, the prevailing trend of separating out information from people has emerged, in the commercial world especially, in moves to elicit knowledge and then store it for access by others. There is a competency gap in many organisations, which the facility to communicate knowledge might be able to fill. In particular the sharing of knowledge is seen as the way for organisations to remain competitive and to get to the leading edge of their industrial sector. But if there is a difference between knowing about something and direct knowledge from using it, can you in fact communicate knowledge?

Can knowledge be shared or traded?

Knowledge is key for organisations. They have realised that much of it is no longer instantiated into technical equipment, as was the case for an industrial plant. Now it resides in people's brains, their programming skills, their abilities to make decisions, to set strategies for the company, to come up with new ideas – and to do so faster than their competitors. For this reason, many organisations have been formalising the process of knowledge-sharing – some appointing Chief Knowledge Officers. Their task is to ensure that knowledge is elicited, stored and accessed. It has not being going well. Sometimes it has been attempted in the context of a merger: two management companies, for example, might need to exchange their knowledge in order to synthesise products, but also to justify the costs of the merger. This is perceived as a competitive situation between units in different parts of the world – if you are not sure your unit is going to survive, why hasten that possibility by offering up knowledge for which the management might otherwise keep you? Also, what incentive is there to take you away from other 'productive' work in order to help others?[10] One powerful incentive is money, and so knowledge trading is touted as part of the new 'knowledge economy' – which, it is claimed, will benefit the whole of society.[11]

I helped develop the world's first knowledge-trading exchange.[12] The idea was that someone with knowledge could express it into a digital form (document, video, audio, executable program, spreadsheet, data) and put it up for sale at whatever price they chose. If they thought that having a trusted imprimatur was helpful for sales, they could negotiate with organisations set up to do that, or with famous-name individuals. The actual transactions were accomplished in real time, so that sellers could log in and watch earnings grow each hour – as some of them did. The system handled co-authorship and percentages with third parties by allocating 100 shares to each piece of knowledge. These shares were also tradeable, so you could sell your potential earnings if you wished. The name 'Knowledge Trading Exchange' reflected the fact that it brought buyer and seller together – so that one could sell and the other buy. Such a system would, for the first time, put a market value on knowledge.

There is the oft-repeated story of the desperate industrial-plant management who, after all their staff had failed to get the plant going, called in a plumber to help. The plumber wandered around, took out a hammer and tapped one particular joint; the plant started working again and he left. The management were horrified when he submitted an invoice for $1,010 and so questioned it. The plumber replied that the $10 was for the work of hitting the pipe and the $1,000 for knowing where to hit.[13] His claim was that his knowledge was valuable. We are, of course, not used to paying for knowledge at market rates – we pay instead according to time and professional sector –

so that lawyers, for example, generally cost more per hour than plumbers. Part of the vision of a knowledge society is one in which the market sets the value of knowledge, so it is commodified and traded just as goods are.

The idea of sharing our knowledge by storing it in a permanent and easily retrievable form is 'efficient' in the sense that it is available for use and, moreover, it is not lost to society when we die. It also appears as a handy short cut for everyone going through the same kind of learning as we did. The traditional way to pass on knowledge was by apprenticeship – a costly, time-consuming transfer of practice, know-how and contact-building which were often kept secret in communities of practice (called, in Europe, 'guilds'). With the breakdown of that system, there is often no one to pass knowledge on to. From the individual's point of view, the transfer of knowledge also panders to our need for transcending mortality. Death without passing on any wisdom, particularly in Western society that believes that it does not need the wisdom of age, can be bleak. We may pass on our genes, but that does not necessarily fulfil the same need. It is as if everything we have learned means nothing. Thus, the idea of knowledge-sharing is both a requirement – for society – and a need, for individuals. However, there remain a number of difficulties: the type of knowledge that can be articulated and transmitted and the nature of what is bought. These difficulties suggest that there are not the short cuts that may be expected.

The main difficulty in sharing knowledge is that direct-knowing is contextualised. To identify and extract all the factors makes the task of communication too lengthy to do easily. Knowledge-elicitation techniques have been refined over decades, yet those going round to 'collect' knowledge often claim that people are unable to articulate it. Moreover, knowledge is frequently displayed in working teams, many of which input into a final product or solution, each playing a part. No single team member knows everyone's contribution: they can only describe their own small part. To get round the problems of contextualisation of knowledge, Chief Knowledge Officers are exploring the role of story – relating events by narrative. This is a kind of archiving that begins by asking questions along the lines of 'Do you remember when ...?' and then recording the answers – an activity previously done in more stable groups with an oral tradition.

The second problem is, assuming someone has offered up the knowledge you seek, how do you find it? However decontextualised the knowledge (in distinction to information) might be, it will almost certainly include words and ideas that are peculiar to the context. Something out there must be usable, but where? Perhaps someone in an entirely different culture or in another industrial sector faced the same problem. In 1966, J. C. R. Licklider saw the use of computing power applied not only to the indexing and searching of documents, but to the pre-processing of them in order to make the content usable – pre-digested, as it were.[14] Tim Berners-Lee saw the World Wide Web as just the first step towards this vision: his ultimate goal is of the 'Semantic Web' – where concepts are linked in such a way that people can get to knowledge, information and data expressed in different ways.[15]

Third, although you can sell knowledge, you buy information. Thus far, I have been using the word 'knowledge' because the debates about knowledge-trading and the knowledge society are framed in these terms. But in fact I would like to suggest that although we can *share* some of our knowledge – particularly that associated with deliberated problem-solving – it is not possible to commodify knowledge *purchase* in this way. You can give knowledge away, but what the receiver (or purchaser) gets is information; it only becomes knowledge when it is used and experienced. In traditional religious language, the master knows and the student receives directions and signs for his journey to knowledge (often communicated through story). This is not to devalue what is received: it is none the less valuable for being information – but information is what it is. The knowledge economy pundits appear to imagine that if only an expert loads his knowledge onto a computer and provision is made for access to it, there is an instant short cut to knowledge. I am arguing that we might have a short cut to tips, ideas, approaches or information, but not to knowledge.

During the 1980s, there were heated debates about information and to whom it belonged, in particular about ownership by employers and employees. Was information a part of the person who had asked the right questions – or was it to be extracted as part of the employer's assets? The answer now seems rather academic to us, for we would assume that information and data belong to the employer in any case of doubt or dispute. It is possible that the same kind of debate will occur over knowledge. Will employers value people for their knowledge or will they value the knowledge despite the persons? Will they value those who can articulate non-contextualised ideas over those who act with subject-specific experience and direct knowing? The conceptual separation of knowledge from a knowledgeable person that is happening in many corporations raises some interesting questions about relations as well as the nature of knowledge itself (that is to say, relational as well as epistemological questions). Certainly there is value in treating any articulated knowledge as data, and then processing it to search for generalisable patterns. However, belief in short cuts, as also found in the optimism of early expert systems, is likely to be disappointed.

Information requires attention

Much of the information we process moment by moment takes the form of implicit questioning. When we listen to the news, for example, we monitor it, constantly asking 'is this of interest or concern to me?' Psychologists tell us that we listen to both news and advertisements in this way. Rather than listening out positively for bits that may be of interest, we filter out what is not (and chide ourselves if we miss anything we wanted to hear about). Although we find and make use of relatively little, we have actively to attend to and filter it all.

The activities of information-monitoring and seeking are not optional, but necessities. In a consumer society, lives are not sustained directly (as in subsistence economies) so people have to attend in order to satisfy basic human needs – food, shelter, utilities, transport, etc. In a commodity-intensive society, people have to be able to get information about the commodities they need in order to select – and this includes determining what is safe and what is affordable. As a result, there are many different routes people can use to search for, monitor and receive the necessary information. A consumer society is one full of surface information and this raises questions about our human capability of handling the large amount of filtering required[16] – and the likelihood of success in finding what we want in it all. The huge increase in the quantity of words and signs has given rise to the expression 'data-smog'.[17] The scale of the increase can best be illustrated by a few historical facts. A weekday newspaper in this century has as much content as the average person in the seventeenth century would encounter in a lifetime. In 1971, the average American encountered 560 advertising messages per day, but by 1997 that figure had risen to 3,000. In 1999, the average British working day contains 171 messages, 46 phone-calls, 22 emails, 15 internal memos and 19 items of external post.[18]

Media advertising specialises in attention-grabbing and holding, because the measure of success is in the degree of attention paid. In TV advertising, for example, success would be measured in terms of the number of viewers and the brand-name recognition achieved (leading to automatic choice of the familiar brand at the point of purchase). The factors for 'success' are fairly well defined: that messages produce an emotional response (achieved especially through drama, suspense or sex), that messages are perceived as coming from a trustworthy source, and that messages are concise (so you can repeat the tagline as an inner liturgy – 'probably the best lager in the world', for example). For all messages, however, the most effective trick is to personalise – then they'll pay attention.[19]

When I was a student at university, I was walking into town one day when I heard a police patrol car call my name over its speakers and order me to stand still. Did I? You bet! As it turned out, the policeman was the husband of the person who cleaned my room. He had recognised me and just thought he'd say hello and have a bit of fun – he kindly explained as my heart rate slowed down again. Personalised messages communicate.

IT greatly facilitates the 'personalisation' of messages in many media – print, email, WAP phones, digital TV and radio. One can imagine all sorts of possible future developments in the personalisation of advertising to catch your attention. Take the highway advertising hoardings: these could display an ad relevant to the car stopped at the traffic lights by reading its number plate, identifying the person likely to be the driver, calling up their current lifestyle profile and then flashing up 'David – buy the latest waterproof hi-fi for your boat' – or some other enticement. Or take mobile phones: there are experiments with superstore messaging to WAP-enabled phones to tell the shopper what and where the special offers are. By using information from reward cards, the offers could be personalised to keep tempting you back.

On the Web, there is an explosion of information with an estimated seven million web pages being added every week in 2000.[20] When using the Web, we filter information in a similar way to listening to news, using the Web pages as signs and pointers to try to ensure that we don't miss what we seek, or what we serendipitously discover we want. Informing here is an active process – we discard and we select. It is this activity of discard and selection that makes using the Web so compelling. We go online to get a single item of information – a task we imagine will take a few minutes at most – and then discover we have spent half an hour or longer. The activity keeps us hooked and closed off to whatever else is going on; even though we are not required to be fully engaged with the content, our attention is grabbed and maintained. This is sometimes called the 'stage' or 'drama' effect: there is so much going on that people get hooked into thinking that the drama they see is the whole world, shutting off the wings and wires that limit, facilitate or enable the interaction to take place.

Whether online or off, we are bombarded with information. No wonder then, that the corollary of the 'information society' is the 'attention economy'. This is a popular term in current business parlance and marketing wisdom says that the future lies in the capture of consumers' attention. Yet for most people there already seem to be too many demands on our attention to be handled satisfactorily.[21]

Front-of-mind stress, back-of-mind creativity

What are the effects and implications of demands for attention for organisations and for individuals? Attending is a 'zero-sum activity'; in other words if you are attending to one thing, you cannot attend to another. Even with the fragmentation of attention (which we consider in Chapter 6), we cannot create any more of it to go round – only routes to avoid giving attention to something.

The art of attention's capture has been well practised by magicians, theatre and other media for millennia. Attention as an activity has been the study of psychologists for decades, as they seek to understand how the mind works, how to operate most productively and how to design jobs, particularly for safety.[22] It is only now, however, that people's capacity for attention is being recognised more generally as a limiting resource at work and in consumer behaviour.[23] This has been precipitated largely because corporations cannot get their employees to attend sufficiently to their goals. In business, people complain of constant change and new projects. With a limited amount of attention at an individual's disposal, some will inevitably be supported more than others.

> The most important implication of the zero-sum rule for managers at traditional organizations is that they need to limit the number of internal programs that compete for employee's attention. We can't overstate the importance

of this point. One Harvard professor's informal survey suggests that the average company has more than 16 major initiatives under way at any given time . . .[24]

As well as the implications for the schedule of work, the demand for attention creates other significant problems for employees and organisations. The activity of attending is a major stress-factor – and this is a cause of growing concern: 'people who keep everything in narrowly-focused, front-of-mind attention are easily overwhelmed, overworked and over-wrought.'[25] Employees cannot operate well when paying close, conscious attention to everything. In addition, the mechanism of response to stress operates in the deep-brain, and the effects range from impairing other functions – for example night vision and other kinds of vision[26] – to narrowing the focus of attention. The demand on front-of-mind attention can therefore become a negative spiral of increasing stress and increasing front-of-mind attention. The visible result is people rushing around, oblivious to all else, getting more and more stressed.

Back-of-mind attention, by contrast, is the more relaxed state that allows us to take in a broader sweep of cues from our environment and allows our unconscious mind to come up with patterns and insights. This 'back-of-mind attention' is both feared and desired in organisations: feared, because it looks so unproductive (Einstein used to spend hours staring out of his office window) – and desired because it is the most effective way to come up with the creative ideas that are required to keep the organisation competitive. Employees have to be released from the demands of tightly focused attention in order to provide the right relaxed, playful, guessing and imaginative environment to function in this way – with no demands on performance at those points.[27] Only then can the unconscious mind be allowed to do its work of picking up hints and cues around and then using them to form patterns and insights. In one organisation where I worked, I bought a group of program-mers a Sony Playstation to have in their office. Many thought that I was rewarding highly paid people with time off – but in fact the team increased productivity, not because they thought I supported and liked them (which I did anyway), but because they had the opportunity of a different mental state in which to muse over programming problems they had encountered. In addition, the team games broke down any tiny animosities of the sort that develop within groups and allowed people to suggest 'silly' ideas to see if the others thought they might work. They proceeded to implement many features we had sought to deliver in our service and which we had previously been told were technically impossible! Optimal mental functioning comes when both concentrated front-of-mind and more diffused back-of-mind attention are available to us and we can move back and forth between them.[28]

Stress, of course, is not something to be shunned entirely. Employees are most productive when they feel the right amount of stress (aversive attention) and reward (attractive attention).[29] The problem is that IT enables much more monitoring to occur and many jobs are now designed to be highly aversive (high penalties for not spotting things going wrong), and demand a large amount of front-of-mind attention (high levels of non-stop concentration). Both are conditions which encourage increasing narrow-focus, tunnel vision and prevent people from being aware of what is going on around them. They tend to select and focus on the aspects of the situation which they judge to be the crucial ones. This spotlight-approach may work sometimes but at other times the wrong task is being attended to.[30] A broad diffuse attention is what is needed to pick up cues not directly communicated to the conscious mind and to develop knowledge, but anxiety, stress and focused attention actually cut this out.

Safety measures and beyond

Information and knowledge are often talked about as if they were in some sense discrete and identifiable objects that are passed from one person to another. As I have sought to show, information and knowledge both require attention and human processing, even if the process is aided by some computer-based data analysis. The widespread belief in the correctness of the data thrown up on computer screen illustrates that many are blinded to all the human steps of attention and inattention, of correct or incorrect processing, of the writing of code and maintenance of machinery that leads to what is on the screen. Reported examples of errors and mistakes abound, of data used in the wrong context, of theft and misappropriation of databases. Our local government council stopped collecting garbage from a street and, when questioned, said that it was not one of their streets – they had checked it on the computer. Eventually after lots of correspondence, they admitted that in changing computer systems the street had simply been left off. It was fortunate that this matter affected health directly, for situations seem to be taken more seriously if they do. There *are* laws, but they only apply in some regions and were introduced to protect trading, rather than the individual directly. Sanctions for poor data entry and manipulation are weak compared to the distress possible, with the entire onus on the person involved to prove a case rather than on the perpetrator. Moreover they are partial in their focus on the end result – on data rather than on the processing. In effect, the persons involved in the data often act as corporations' quality assurance process. As society moves to greater dependence on the computer and data, the equivalent to the safety measures of products, such as cars, and of processes such as drug development, might be appropriate.

The activities of information-processing and knowledge-acquisition take time and attention. Both are necessary for surviving in a society where people depend on purchasing commodities and where labour is given as a commodity. However, time and attention are under siege from many competing sources at work and in living. IT and the 'presences' of the active hardware of cyberspace exacerbate the problem of finding sufficient space to process data into information and knowledge. In particular, the knowledge of the kind that supports being flexible, managing change and the creation of new ideas, is direct knowledge and needs space for pondering and reflection.

The law already recognises the limitations of attention, especially front-of-mind concentration, and that it is a zero-sum activity. At work, the hours worked in continuous monitoring are circumscribed by European law, with compulsory breaks and maximum working periods. While driving in Britain, the police can pull you off the highway and charge you with an offence for attempting to do two things at once – for example, if they see you attempting to drive with an arm around your spouse or while talking into a mobile phone. I was nearly run over by an out-of-control car when the driver was simul-taneously trying to turn a corner and argue on the phone with gesticulating hands. The law also notes our weakness in what attracts our attention: advertising laws and codes of practice limit the content used in competition for our attention, demanding the avoidance of too much sex, drama and sus-pense, even though these are legal in other contexts. Still other laws limit the attention paid to one, making over-attention to a person illegal unless it is reciprocally accepted. Filmstars, sportstars, politicians and news-presenters, for example, require attention to keep viewers and supporters; however, they feel threatened by too much attention, as when they are stalked – even if completely harmlessly. (In many mammals, attention is considered as threat, both by predators and prey, which is the reason why cats don't like being stared at.) These laws are for safety, rather than well-being, but illustrate that there is an underlying ethic on behalf of persons for their health and safety. Consideration of ethics in this area may be further pursued as the issues of the extent to which attention can be divided and the amount given are explored over the next decades.

As I will argue in Chapter 7, such issues are able to be tackled by forming groups of people with a common concern who can engage with the powers-that-be from a position of lived experience.

chapter five

Patterns of Relating

IT, like any new means of communication, brings with it changing patterns of relationship. We look again at the patterns of relationship we know, so as to learn to focus on what is valuable in the new ways of relating.
- What effect does the Internet have on friendship?
- Is an Internet community a real possibility?
- How do online encounters compare with face-to-face?
- In the global world of cyberspace, who is my neighbour?
- What are some of the legal implications?

NEW MEANS OF communicating remodel the existing patterns of relating. How we do it, when we do it and with whom, are all changed as we adapt to new technologies. To focus on the new modes of communication in isolation can be misleading: rare are the times when a new technology simply replaces the old – for example, it usually creates contacts that would not have been there before. But rather than asking if the new replaces something of the old, or if there are changes to particular relationships, it is better to ask about the patterns of relating. It is the whole pattern that changes, rather than one particular facet.

Any new capability poses questions about what we did previously. When people say, 'you can't have the same quality of conversation over Internet', I want to ask – is that really true? What were the significant features and qualities of our conversations using 'old' technologies? How many of those conversations could not be carried out as effectively online? I would suggest, for example, that many of our phone conversations (at least those that ask for information, receive information or give instructions) could be carried out equally well over Internet – and often with additional benefits such as fewer interruptions to work, or indeed to leisure.[1] As existing ways of relating are re-examined and then seen in the light of the new, we may learn more about what is important in them and start focusing on what we value. The process of being assimilated into and altering existing patterns of relating is not unique to Internet technology; writing, the postal service, the telephone and mobile phone have each gone through such an initiation.

Our first focus on relating is not therefore 'ethical' in any usual meaning of the term, but concerns those relationships, how we are to value the communications within relationships and what impact the new may have on them. We begin by examining friendship: does the Internet make it easier to have friends through using email and chatrooms or harder because of not meeting face to face?

This leads us into the same question for communities: can you have an Internet community? In relating to others, does the medium help or hinder? People all comment on the lack of visual and bodily cues, but how does this affect communication? There have been studies researching this, mainly in the context of work, and we briefly look at the findings to discover what is valuable about online and about face-to-face relating. One of the key questions that people ask relates to neighbours – who is my neighbour in this new life-world formed by cyberspace?

Ethics, in the more usual sense, *are* involved, for example in the transmission of messages: did the whole of the message arrive (with all attachments etc.)? Was it 'safe', or could anyone have tampered with the content? What rules apply to the content? Is it considered personal or public? Who owns the right to copy and forward it? In the final section we will consider some of the legal implications of the new forms of communication.

Changing friendship patterns

Internet communication does not seem to change the number of our good friends; it does, however, change who they are. Most people have an average of around thirteen friends and this has not changed since research began in the 1930s. Friendships are built on contact, but this need not be exclusively face to face, and need not have been initiated by meeting in this way. Indeed, first meetings happen in every way imaginable, but in continuing the relationship, three things usually have to happen: frequency of contact increases, content of discussion broadens and a wider variety of media is used to communicate. Contact frequency normally grows to a point where each person in the friendship knows what to expect from the other – then either it remains at that level or diminishes again. The content broadens: perhaps one person risks sharing the names of their family and the other reciprocates; perhaps they discover a shared interest and then take part in it together. Friendship grows through broadening the topics covered until each is comfortable with speaking what is on their mind. The channels used to communicate increase in number: perhaps a card is sent after meeting, followed by a phone conversation; or perhaps two people met over the Internet and then decide to meet face to face. Whichever mode of communication is the starting point, the next step is always a risk.[2] This is because you may be more or less competent in the new media, and in any case, what you can express in different media is different. If you normally talk to a particular person on the phone, they might never have seen your handwriting. Having articulated thoughts to someone online, you might wonder if it will be as easy to talk face to face. You don't know until you try, so people 'test' their friendships by extending the means of communication. Friendship grows as contacts increase in frequency, crossing media, and as the content broadens. Early research in online communities found that scientists, for example, were able to convert remote acquaintances to friends much more easily by having computer conferencing (both public and private, one-to-one) – which provided a convenient, additional route by which to expand topics of conversation and increase frequency of contact.[3]

Internet communication, including email and 'chat', enables some relationships to be created and maintained that would not probably survive otherwise. Characteristically, these friendships settle into a pattern that many find is easier to maintain using asynchronous communication – probably in addition to other channels of contact.[4] Because telecommunication enables us to retain some friendships that might otherwise be replaced by new ones, by implication we are less likely to make friends with those we meet face to face. Friendship patterns have therefore been changed by the new technology.

Some friendship patterns transfer far better into Internet communication than others. In general, friendship has different functions for men and for women. In addition, for men there are notable differences between working (blue-collar) and middle (white-collar) class.[5] Blue-collar men form friendships around reciprocal help and action, much in the way tribal economies work; whereas white-collar men talk more and find ways of understanding and reframing the world through each other's friendship. Women, on the other hand, seek listening ears and sharing of life's difficulties – in other words, emotional support. I note these generalisations only to indicate that friendship content may transfer to a greater or less extent onto the Internet and that no global view can be taken of what might be transferred.

Can there be online communities?

Communal life in the Western world is disintegrating, so it is claimed. Many feel that the quality of community life is deteriorating. They see public spaces left untended or vandalised, a loss of respect for the personal property of others and a smaller proportion of the populace actively engaged in local politics. The retreat from community life is illustrated by the change in urban architecture – particularly visible in London, England. Victorian houses had their family 'living' room at the front of the house, connected to public space; now architects and builders are engaged in turning houses back to front, so that living spaces are sequestered into private areas at the back. One tenet of popular belief says that this trend is a direct result of the entertainment and communication services removing the need to go outside to meet in public spaces to hear and exchange news and to state and form opinions. This, they claim, is one sign of the breakdown of community life. But if people are not meeting and forming community in local public spaces, then perhaps they are through Internet services themselves? Is it really possible to have a so-called 'virtual community'?

Researchers and scholars who describe groups interacting online as 'communities' have frequently been criticised and challenged as being too ready to subscribe to utopian dreams of a new society. They are 'sham' communities, the objectors proclaim, because what happens between people in cyberspace is a pale imitation of the rich texture of interconnection in a 'true community'.

Some people do not want to accept that Internet communities either exist or could ever be possible. Why is this? One reason is that the Internet model of community is built on word-based communication and shared interest, which is rejected as a basis for 'true' community. In other words, the activities of an online community might actually undermine the concept of community itself. This is the concern of downsayers such as Neil Postman, whose particular criticism of Internet communities is that they lack any sense of common obligation.[6]

We will discuss these ideas, taking into account some definitions of community and some evidence from the Internet. In doing so, I will deliberately avoid the word 'virtual' since it has a rather mixed set of associations and can be misleading.

In any case, we should bear in mind that communities on the Internet are groups of real people who meet and have real relationships sustained through the Internet – people with emotions, beliefs, loyalties, concerns over appropriate behaviour and how to deal with newcomers in their midst. Other expressions in common usage include 'online communities', 'computer-mediated communities' (CMCs) and 'cyber-communities'. I prefer and will use 'Internet communities'.

The most prevalent understanding of the word 'community' appears to be one associated with a group of people living in a particular geographical space. Historically, however, this is not the only signification. Introduced into English in the fourteenth century, the meanings of the word 'community' coalesce around two ideas: social grouping and quality of relationship. The application of these two ideas has ranged over a number of senses: the common people as distinguished from those of rank; a state or organised society; the people of a district; the quality of holding something in common; and a sense of common identity.[7] Raymond Williams describes it thus:

> The complexity of community thus relates to the difficult interaction between the tendencies ... on the one hand the sense of direct common concern; on the other hand the materialization of various forms of common organization, which may or may not adequately express this ... What is most important perhaps, is that unlike all other terms of social organization (state, nation, society, etc.) it seems never to be used unfavourably, and never to be given any positive opposing or distinguishing term.

Over the past few decades the word has been increasingly applied to particular special interest groups; for example, we speak of the 'gay community', the 'business community', or the 'disabled community'. Herein a number of ideas are implied – the sharing of a common lifestyle, having common concerns and interests and working together in political lobbying and representation in policy-making.

The fact that the word has so many different meanings

attached, means that debates on whether or not communities can form over the Internet can be in trouble from the start. In preference to making any attempt to discover and defend any definition of community, we can find a useful alternative suggested by the philosopher, G. Graham. He gives three conditions to apply to a conceptualisation of 'community' and thereby distinguishes between three different groups of people:[8]

- *subjective interest*: a group of people who are interested in the same things;
- *objective interest*: a group of people who are, as a matter of observable fact, affected, beneficially or adversely, by the same things;
- *defining authority*: a group of people whose common identity as a community is by their owing allegiance to a mutually recognised authority.

The first two of these conditions are self-explanatory. Stamp-collecting and sailing would be examples of subjective interest groups, while farming and stock trading would be objective interest groups. The third condition needs teasing out a little for clarity. Deriving this condition from monastic communities, Graham stipulates that the members join voluntarily but are then subject to a rule that determines both what their objective interests are and what their subjective interests ought to be.

Graham examines evidence to consider whether Internet groups could ever constitute communities under his three conditions. He concludes that ' ... there is no reason in principle why an Internet community should not have the same essential features as an order of nuns.'[9] It so happened that the groups he was able to examine either did not fit his conditions (he decided) or did not yield sufficient information to make a judgement either way. However, Watson gives an example that I believe would qualify.[10]

Phish.net was an online fan community, about fifty thousand strong. Fans voluntarily joined as members of Phish.net. Once part of the membership, their subjective interest (the band) was guided by the norms of fanship – with the threat of being 'flamed' and rebuked if they did not support the aims of the group. Their objective interest was subject to factors which affected them all, such as the appearance (or non-appearance) of the band; their identity as fans was maintained by going to the concerts and swapping recordings made there. There were factors which were perceived as affecting whether the band played or not and at which venues. For example, in several cases, bad behaviour at concerts might have led to cancelled concerts so the need arose to formulate behavioural norms. In one case, the perpetrator of a 'rush' to get in free and a user of fake tickets apologised online, saying he did not understand the wider implications of his actions and wouldn't be doing it again. The online fan club and band even worked together to issue leaflets about behaviour. This Internet group clearly satisfied the first two criteria; did it satisfy the third? People gave allegiance to the group and developed norms and values as a result of 'a repeated and continued need to affirm the foundational values of the community in the face of threats to those values'.[11] This would not be very compelling if the norms and values were a rearguard reaction against decline, but in fact they were the result of rapid growth. The group would therefore, in Graham's terms, qualify as a community:

> Any such group which can be said to have the basic values of a community is one to which we belong voluntarily, but not merely in virtue of our interests alone, whether subjective or material. Rather we belong because we accept and adhere to norms and standards (and are required by others to accept and adhere to them) which define and constitute membership, and we remain (because accepted as) members only so long as this is true.[12]

We have considered Internet communities from the perspective of shared relationships among people – as opposed to shared communication in the same physical space. But cyberspace is still dominated by word-based communication and there is distinct lack of visual and other non-verbal cues. What kinds of relationship are possible if you are confined to words?

We now turn to consider those relationships and how Internet communication changes the way we relate. Much of this research has been done in work situations and so the next section focuses on the work aspects of telecommunication versus face to face.

When does wired working work?

Do we need to meet? Or can we work 'over the wires'? What can be transmitted successfully? And what might get lost? A number of research studies have been done in this area, comparing both telephones (remote audio) and computer-conferencing with face-to-face. Some things, it transpires, are the same or even improved online, others are better face to face.

What we might call 'rational' discourse is not affected by the type of media used for communication, whether word-based, voice-based or including a visual element (this was true even in experiments using only vision and no words). Problem-solving is in fact improved by means of telecommunication, whether Internet or voice or video-conferencing. This is because among a participating group more people feel able to contribute and to express themselves freely, so a wider range of ideas is offered.

The situation changes, however, when the objectives themselves need to be discussed, or when judging and resolving conflicts of opinion, or when trying to reach consensus and support for the decision made. Without visual cues, it is difficult to judge someone else's opinion on a particular subject.[13] The voice also plays a key role in negotiating objectives: it communicates emotion and without it people find it hard to handle conflicts (we will discuss this further later in this section). Reaching agreement therefore is best done face to face, especially on complex decisions. Work on comparing computer conferencing with face-to-face found that reaching agreement involves much use of non-verbal cues, changes of position and moving towards supporting an authority figure or towards a person offering a particular solution.[14]

But, some argue, there will be continual technical improvements and eventually, as a generation grows up being familiar with this technology, there won't be the same difficulties. It is certainly true that many of the major technologies have only been fully adopted when the generation alive during its invention has died and the next one starts to appropriate it as part of everyday life. There is, however, little evidence to suggest that the basic cues developed through voice and movement (body and eye) have adapted for technologies that have been around for longer than telecommunication. I therefore believe that the needs and abilities of humans in the field of communication are more fundamental than this and not necessarily adaptable to any prevailing technology. One leading researcher who works on developing visual cues agrees with this view, considering that remote communication will always serve 'rational' communication better and so we should look to it for supplementing, not replacing other forms of communications.[15]

Internet communities have found various ways to compensate for the shortage of visual and verbal cues. In Phish.net, authority was asserted by members presenting large lists of recordings of live concerts, by messages displaying detailed knowledge of the band and by repeated presence online.[16] Other members fell in line with their norms and practices in face of the perceived 'authority'. In email and computer-conferencing, cues have developed to express emotion and gesture. The rules of 'netiquette' and use of facial symbols help to pre-empt misinterpretations of what has been written and temper any negative emotional reactions. A husband and wife with a young child used to argue in computer-conferencing on whose turn it was to collect him from the baby minder, and developed symbols that intimated 'I love you – but it's still your turn.'[17] Such symbols to indicate that what is said is meant kindly or as a joke are now commonly used, for example ':-)' – a smiling face sideways on – (so much so that when I tried to type that in, Microsoft Word immediately interpreted it as a 'smiley'). The third main development to compensate for absence of cues is found in online debates. In a traditional debate (not an online one) a new person entering in the middle can find it difficult to know what has already been said. To help the newcomer (whose arrival is probably obvious) the synthesiser of the group will often do a quick summary. Online, it is different. One might expect that the written record of all the contributions would serve the same job, but the sheer quantity make reading all the history time-consuming and difficult. Two methods are used to overcome the problem: giving occasional summaries in the course of the discussion[18] and developing a separate FAQ (a list of frequently asked questions). The latter is not just a list of answers to questions people ask; it orientates them to the direction of the debate and indicates where the various lines of discussion have got to.[19]

The ability to use and understand any of the cues described does not, however, imply replacement of the deeper human perception and reactive systems. People continue to meet face to face for business to check out whether or not they can trust each other, to hear their opinions and to check that their goals are sufficiently aligned. Only having checked these out, can problems that arise be resolved using telecommunications.

The most significant feature of non-verbal communication, as used for email, in chatrooms and computer-conferencing, is that the emotion is largely removed. Being activated from a deeper part of the brain, feelings affect the voice more than the selection of words.[20] Quite unconsciously, listeners pick up emotions in the voice of a speaker. Even if the speaker wishes to disguise his or her emotions, it is quite hard to control the effects on the voice – for example, depression enhances nasal resonance, shyness softens the amplitude, anxiety introduces breathiness with elevated pitch, joyfulness animates through pitch, and hopelessness removes most of the lower frequencies through which feelings are transmitted. It can also be difficult to distinguish accents and identify people on the phone, because for both we need to hear the higher frequencies that get lost in transmission – even though we can tell their emotional state in a few seconds. Oral gestures, such as rhythm, speed of speaking and degree of articulatory force are used to indicate sincerity, confidence in what the speaker is saying, and humour. Oral gestures are also used to persuade others and to indicate the degree to which one is listening to the other person. The absence of voice and oral gesture in an email explain why it is so easy to mistake the intention of the sender. We can only guess at the 'tone' and if we guess wrongly we risk giving an inappropriate response that can escalate some situations ('flaming').

To summarise, there are good reasons why it is easier to reach agreement face to face: the visual cues give a much clearer picture of the other's opinion and by the vocal cues you can gauge the amount of emotional commitment they have to an idea – you have the essential information to begin a negotiation. Second, you feel you have greater support for working through the outcome when you know you have been listened to. There is a radical difference between posting a message on Internet – not knowing if the recipient or group has read it, and speaking – knowing that you have been heard by a person or group physically present. Being assured that your voice has been heard is a critical part of communication and community. S. G. Jones, taking this idea further, suggests that the lack of being heard is one reason for disaffection with political processes.[21]

Why we need international friends

Many would say that another reason why meeting face to face in business is important is the building of trust. This is done through both individuals and networks. Trust is not only a matter of personal judgement exercised through meeting with the other person and then assessing their actions. Meeting someone is usually only one of a whole set of actions that we take when considering whether or not to do business with him or her. There are formal procedures – looking at their accounts, checking out who their solicitors are, talking with the local

World Trade Centre, checking with commercial advisers on whether they have a clean record and so on. Informal networking is highly used too – 'What's so-and-so like?' we ask conversationally, or perhaps more explicitly, 'I am thinking of placing a large contract with so-and-so, do you know any reason why I shouldn't?' I have even called up former competitors for advice on suppliers in an area with which I was unfamiliar, and received valuable information. In order to use an informal network, you have to have one. International business therefore depends on international personal networking. To find suppliers and project partners and to avoid bad deals, you need to build relationships that will then act as a community which you can trust when you need help.[22] International business draws you into international communities with their need for personal investment, participation and maintenance.

The point here is that globalisation in commerce necessitates building international personal networks which take people away from the more local opportunities. Considering this – together with the conclusion, drawn earlier, that Internet communities are possible (if perhaps not, at the moment, common) and that there is a trend for friendships to be created and maintained over distance instead of being replaced by new, local ones – leads us to ask afresh the question, 'Who is my neighbour?'

Who is my neighbour?

'Neighbour' is a loaded term for the religious and non-religious alike. In everyday parlance it refers to the person living in the next house – and hence has all the connotations of community. If I am a Christian, then it refers to the person whom I am to love as myself. Both have an implicit sense of proximity – of the geographical nearness of the one to be treated as neighbour. Once that geographical locality is transcended by travel and messaging services, the idea of neighbour has to be reviewed.

The Internet introduces another step in the debate because, as we have seen, the distant relationship becomes easier to maintain once the other person does not need to be there at the same time (the communication can be asynchronous). If we really can have online communities then, it is argued, neighbours are 'there', rather than geographically 'here'.

Let us review the possible meanings of the word 'neighbour'. First, it might simply mean a member of a community (and that community could be online). Second, it can mean no more and no less than the person who inhabits the same space. That space could be in the material world, and then it would apply to the people physically around, or it could mean cyberspace, and then it would mean the people 'around' in the places you 'go to'. In both there is little control over who those people are except by choosing your place in the space. Wherever you choose to be, they are there, with all their differences and similarities.

There is a third possible meaning I would like to explore – the one that seems to underlie a phrase such as 'they were real neighbours to us'. This is not about being a member of the same community, nor of co-location in some space, but about helping in our times of need. The proverb 'better a neighbour nearby, than a brother far away'[23] is given in the context of 'when disaster strikes you'. My definition for exploring the concept of neighbour is derived from this: a neighbour is one who is able to be aware of your need and who can help you in your time of need. The definition has two parts – the awareness and the activity of helping. Note again that this person is not chosen. I have deliberately kept in the personalisation of 'you' here. Jesus' story of the 'man who fell among robbers' (otherwise known as The Good Samaritan) is given in response to the question 'who is the neighbour whom I am to love as myself?'[24] The answer is the one who had mercy: the neighbour to the man who fell among thieves is the one who stopped to help. The neighbour is the one who helps in time of need and, irrespective of who they are, they join the group of those we are to love as ourselves. The 'Go and do likewise' at the end of the story must be first taken to mean that the expert-in-the-law is to go and love those that help him in his need, irrespective of faith, background etc.[25] Islam too has a strong story base in the Middle East of welcoming those who help Muslims – the Friends of the Friends of Allah.

We can now ask the question 'who is my neighbour?' of the definition: a neighbour is one who is able to be aware of your need and who can help you in your time of need. There is demand-led and response-led awareness: demand-led, when we request help, and response-led, when it is offered. The Bible records both kinds of awareness – a man wakes up his neighbour with demands for help in the form of bread to feed an unexpected visitor, and another sees a man lying beaten up and robbed by the side of a road. Demand-led awareness, taking the form of a request for help, is more immediately amenable to cyberspace neighbourliness than response-led. It is a help that is in no way diminished in value even if it is also mediated, for example by calling emergency services, or organising nearby people to meet the need. Response-led awareness is of the kind where the neighbour says 'I couldn't help noticing that . . .' or 'I heard that and I have . . .' or comes round to see if you are OK after hearing a scream and a thud. The noticing and attending is clearly much harder to do online (although sometimes gossip can take this role and initiate action). It is therefore possible to be a neighbour online (in demand-led awareness), but less likely because the time we most need help is often the time we are least able to ask for it (through response-led awareness). We can also note that technologies for communicating are used to turn response-led awareness into demand-led. For example I hear people say, 'He's getting old – he should have a mobile so he can call for help or install an alarm circuit so we know if he has fallen.' Technical solutions mean that we don't need to attend to someone until a demand comes. In this sense the tools of IT are being used to reduce the noticing that is part of neighbourliness.[26]

I would like to draw out one other aspect that seems to be implicit in the epithet in '*real* neighbour', and that is the sense of not being abandoned to sort everything out ourselves as autonomous individuals or nuclear families. What we need, therefore, are signs and acts of presence – and these are as important as the activity of help. But if, as a result of communication technology, response-led awareness is making way for demand-led awareness, then signs and acts of presence are less likely to be given: people may wait to make contact until the demand for help comes. It is also the experience of many that others draw away when disaster does strike.[27] So I suggest that the anguish at the heart of the question 'Who is my neighbour' is not directly associated with the effects of cyberspace and now-distant friends and acquaintances, but with a heartfelt concern – 'Who will be there for me in my time of need?' – that arises within a pattern of changed and mediated relationships.

New technology, new legalities

New patterns of relating not only have direct impact on our personal, communal and business life; they also have legal implications: for example, who owns the content and who has the right to make copies? What kind of surety can we have that our communications will arrive without interference and without examination?

When sending letters through the postal service, the copyright remains with the author, but the recipient is owner of the letter. When Major James Hewitt revealed the content of the letters he owned sent by Princess Diana, he broke copyright law by not seeking her permission. In the case of telephone calls, both parties must agree and give permission before any recording is made. Each new communications channel raises a new set of questions over ownership and copyright of content. What constitutes law and practice with email, for example, is still being worked through. I have found that older people using email will usually write to me asking for my kind permission before forwarding my message on to someone else – treating it like a letter. Others just forward it on anyway, as if being the recipient gives them the right to copy. If this trend becomes the general practice, then indeed every email may be considered as in the public domain in the sense that the sender will have no control over who will see it. In contrast, in one long-standing online community (the WELL community), the writers own their words (both copyright and ownership).

Currently, recipients of email have no way of verifying the identity of the sender or the content unless encryption has been used. The UK government is considering making digital signatures legal as verification of emails – by which we mean not the digitised graphic image of a signature (sometimes used in word-processed letters and email), but a secure method of proving that a message could only have come from the sender.

The situation is slightly clearer in the context of employment. Most employers have contracts which include regulations on the use of the employer's equipment and resources, and which require rights of access to all communications where the employee is representing the company, either directly or indirectly (as in trade parties or professional conferences). The corporation may have to defend itself in lawsuits years after an employee has left, and so records of what was said by and to whom can be vital. In essence, every email sent by an employee is considered as company property. In practice, of course, most employers allow some essential personal communications on the phone, by letter or email. That leaves one area – union affairs, where, by UK law, unions have the right to go about their business without interference from management. Do they have a safe space to do so where management cannot read all email? In cases of concern, new safe spaces for such activities can be found – for example by using webmail with email stored on distant non-company computers.

Two aspects of email make inspection easy – first it is digitised and so easily searchable for key words or phrases. Second, email works with a store-and-forward system, so the server back-ups automatically capture almost all email passed. It is easy to go back and find emails, even though both sender and recipient might think that they have deleted the communication – as some employees have found out to their cost. In fact this aspect of electronic storage is increasingly used by companies as their archiving device, requiring employees to note and file communications with outside agencies, and to communicate internally what they have done. The state may demand the right to monitor communication activities if they have reason to suspect that citizens or aliens may be undermining law and order. We consider the UK's Regulation of Investigatory Powers Act 2000 in Chapter 7.

Patterns of relating are one of the most common topics of conversation when the subject of cyberspace is raised. People want to know what is possible and what might occur. They are faced directly with competing views – of teenagers shut in their rooms on the Net (often boys), barely talking except to grunt, at the same time as parents are paying expensive mobile phone bills for children who never seem to stop communicating with their friends, and are also experiencing dislocation of familiar communication patterns at work and as consumers. How do you make sense of what is happening?

IT and cyberspace are undoubtedly changing the patterns, primarily by enabling the maintenance of distant relationships among friends, communities, working groups and away from co-located neighbours, colleagues and those around that might become friends. The primary fear I have identified is that of wondering who would be there for a person in time of need. From research we have learned of the richness of face-to-face communication in understanding the opinions of others (visual) and perceiving their emotions (aural); these are essential for building relationships that have to agree objectives and be maintained during difficult times. We have also learned of the value of electronic communication for problem-solving and sharing ideas, as well as the routine support of 'rational' communications, such as organising and arranging. We considered how the communication is increasingly experienced as demand-led and not response-led, as the devices of cyberspace intrude or are expected to be used – 'you didn't ask for help', people say accusingly.

Although there are ethical and legal questions, which are not to be minimised and some of which we have touched upon, the primary question is 'what kind of patterns of relating do you want?' I will explain (in Chapter 8) why the answer for me includes considerable face-to-face relating in a risky exposure in order to develop new kinds of relationship in a community of faith. How we choose to use or purchase channels to contact friends and relatives and work colleagues will help create new patterns of relating – patterns that are undergirded by technical means that strengthen or weaken the technological institutions that support them, and so determine what is considered 'normal' in society.

But much more happens in communication than its success or otherwise, its influence on the channels and the formation of new patterns. There is also the aspect of what we are for others, and what the situations and presence of others, whether distant or near, elicit from us. This question is about who we are, our identity and formation of self, and we turn to it in the next chapter.

chapter six

Identity and Self

How is the formation of 'self' affected by IT? Does our changing society lead to the creation of multiplicitous selves? What are the likely consequences of experimenting with identity in cyberspace? What happens when our data-image is used by organisations as though the collection of information about us constitutes our identity? Finally, we look at how privacy is affected by the development of cyberspace.

OUR UNDERSTANDING OF self and identity is apprehended through interaction with other people: through others we learn who we are. I have learned that I am enthusiastic because others react to me as if I am, that I am intelligent because they tell me so, that I listen because they talk and tell me things others don't get to hear. I know those things now, but there was a time when I was not so sure: it took a gradual accretion of experience and a process of sifting to find out who I was. Sometimes in a British culture, where there is so much low self-esteem because no one affirms your strengths, the only way to gain this sense of identity is to ask others for their perceptions, and many are finding this an essential step in valuing themselves. By interaction, people come to know their identities.

My identity is bound up with what I give to others: what their needs call into being within me. I am enthusiastic, for example, because that is what I am for others. I have appropriated this characteristic as a meaning in my life and a guide to my behaviour. I experience many people as tired and weary – my enthusiasm lifts them in renewed hope. In a world where many have difficulty finding others with time to attend to them, my attentive listening helps them to make connections and renew direction in life. In this sense, I see inspiring others with hope and helping them to find direction as one part of my life-meaning. My identity isn't only those sorts of characteristics. It also includes the way that I tend to interact with people (as revealed, for example, by Myers-Briggs typology); the things that trigger and sustain my interest and action (as analysed by SIMA), or my deep motivations for acting one way rather than another (as uncovered by the Enneagram). A major defining part of my identity has to do with what I think I am here for – and for me that is to develop a meaningful technology, and experience of that technology for others.

The formation of identity, as we begin to see, is the product of a highly complex combination of interconnecting factors. There is one other aspect, however, that is significant when we come to consider the issue in relation to cyberspace. This is the continual (re-)negotiation of who I am as I engage with different people in different areas of my life. The great number and variety of situations and people that I meet requires such continual and rapid switching of interaction, that I sometimes feel I am being fragmented.

It will be evident that 'who am I?' is an important question that keeps coming up in the context of the IT and cyberspace. It is not a theoretical question, because my view of myself alters what I believe I can do and accomplish: it can be liberating or limiting. How we view ourselves is closely tied with what we can do, where we can travel, what information is accessible. What we believe about our capabilities, including access to technological help, is constructed at a deep psychological level. For example, Hippocrates noted that people with leg injuries, even if healed, frequently did not walk again because they had constructed images of themselves as non-walkers. It remains true today that the limitation of mobility resulting from leg injuries changes one's perception of one's capability.[1] P. Virilio, the French new technology commentator, in the midst of a generally rather pessimistic forecast, suggests that a similar limitation may occur in an individual who becomes 'terminal man': he (or she) sees the world through the window of the screen and, seated and static, neglects what can be discovered by walking or meeting people face to face.[2]

In this very sparse and selective preamble on the theme of identity I have tried to pick out some of the key aspects relating to the role of information technology and cyberspace in identity formation that we will consider in this chapter.

Historically, there has been a close tie between identity and work. But work is changing and this means rethinking how we derive the view we have of ourselves. Old patterns of stability in work and social life are giving way to new ones of adaptability and flexibility. Established communal activities and symbols that used to help maintain stability – and thus make it easier to find identity – have given way to new, more fluid groupings and temporary surfaces, which require us to find new ways of creating identity. In the first part of this chapter, we explore the ways in which the formation of 'self' is affected by information technology, and the new paths to identity that are emerging.

Throughout history, people have been interested in the idea of exploring – or even creating – different or alternative identities. Cyberspace certainly opens up a whole world of new opportunity to escape all the old groupings and symbols that defined us, to present new images of ourselves and to engage in fantasy play with alternative selves, either in fun or as a kind of virtual therapy. In the second section we look at the question of whether the changing society leads to the creation of a multiplicitous self and at the possible effects of experimentation with identity in cyberspace.

Some may feel that cyberspace offers an exciting prospect of liberation from one particular identity but, ironically, it is also the place where a fixed identity is created. The trail of data we leave in our wake as we go about our daily lives is accumulated into a 'data-image' that is used by organisations to make life-choices for us. As far as they are concerned, that data-image is our identity. In the third section, we examine the implications of data-images and suggest some possible responses.

This leads us to the subject of privacy, and in the final section we reconsider privacy in the context of identity and boundaries.

Work and consumerism as source of identity

Juggling is a common metaphor for maintaining continual action in a number of projects, attempting to apportion attention so that none loses momentum or direction. The new technologies enable us to be 'present' to a variety of projects 'simultaneously' – by means of email, mobile phone, video-conferencing etc. – even if the projects are scattered across the globe. In this context, I can then be described as a 'distributed self' as I attend to projects in different organisations and around the world. I might feel this 'distribution' of self nowhere more acutely than when I cross the private–public divide, perhaps reorganising, minute by minute, the collection of kids from school or the buying of food in the midst of the continuing work projects. Although such multi-tasking has always been practised, traditionally more among women than men, IT facilitates it and even sets up the expectations of it, as people are encouraged to employ the new tools for this very purpose.

Many tasks are performed within particular roles – that is, the different functions we fulfil within the norms structured by the institutions and organisations of society. We all play many different roles through the course of life: employee, football player, churchgoer, political party activist, parent and, certainly, child. Despite the large number of possibilities, society has never been comfortable with the idea of a person having multiple *public* roles. We like to know where we are with people, Jung observed.

> Society expects, and indeed must expect, every individual
> to play the part assigned to him as perfectly as possible,
> so that a man who is a parson . . . must at all times . . . play
> the role of a parson in a flawless manner. Society demands
> this as a kind of surety: each must stand at his post, here
> a cobbler, there a poet. No man is expected to be both . . .
> that would be 'odd'. Such a man would be 'different' from
> other people, not quite reliable. In the academic world he
> would be a dilettante, in politics an 'unpredictable' quan-
> tity, in religion a free-thinker – in short, he would always
> be suspected of unreliability and incompetence, because
> society is persuaded that only the cobbler who is not a
> poet can supply workmanlike shoes.[3]

As hinted earlier, technology's effect on multi-tasking is that
roles that were previously discrete often intrude upon each
other. The idea of playing out a coherent, compartmentalised
role, as in Jung's vision, is no longer easily possible; other areas
of life keep elbowing their way in. Moreover, the work role
itself is changing.

Work was one of the primary ways of acting in and relating
to society, and so associated with identity-construction. Luther,
in the early sixteenth century, did much to relate work and
identity when he cast the work role as the secondary vocation.
The first was to be a disciple of Jesus Christ, and the second to
express that life as service through a work role (which specifically included the work of being a mother or housewife). He
argued forcibly against the idea, prevalent at the time, that
only priests, monks and nuns could be considered as 'full-time'
Christians. Those times have now passed, but the religious idea
of role being tied up with vocation is still strong – too strong,
some argue, suggesting that we should now drop the idea of a
second vocation entirely.[4] The idea is an inappropriate one
because of new patterns of employment and work: frequent
changes in jobs, 'portfolio' working, women taking a break
to have children and then returning, skilled workers finding
themselves redundant or that their industry has vanished, and

so on. When identity is associated with work, people in any of these positions will be prone to identity crises. There has to be another way to find identity.[5]

In fact, work as the source of identity is already being dis placed by the dominant structure of living in Western society – consumerism. Consumerism is not only about what we buy in shops (or the online equivalent) but a whole set of choices of lifestyle including house locality, schooling, the structure of a friendship base and church – as well as work. The cultural space that was occupied by work in forming identity is now filled by consumerism – so that increasingly, new acquaintances don't ask 'What do you do?' so much as 'What are you doing at the moment?' or 'Where do you live?' in order to determine who you are. One analyst says that the future growth of con sumerism depends on having each person believe they are able, individually, to choose their identity and to change it at will.[6] Without the ability to make people think 'Today I will be a Goth', or 'Tomorrow I think I will identify and act as a Native American', corporations cannot sell enough to keep expanding their markets. Continual expansion requires the possibility and practice of changing identities, all of which are carefully moni tored by IT and the latest fads advertised by media. This is a vision of a life characterised not only by multi-tasking but by multiple selves.

New ideas of self

The postmodern hypothesis suggests a chosen and changing idea of self, one based on shopping for an identity within consumerism.[7] This fits particularly well with the 'experience economy', where we respond to and buy experience, rather than goods or services. The ultimate aim is to 'take us out of ourselves'.[8] Lyon calls this kind of being a 'plastic self', desirous and willing to be taken out of who we are in order that we accumulate experience and can try out different lifestyles. At the same time, he identifies the emergence of another kind of self – the 'expressive self' that still seeks significance and tries to create some authenticity of identity. The identity in this case is based on the well-being of the 'voice within' – 'I feel good about this so it must be me.' The key element of the 'expressive self' is that the person retains some sense of their own story, but is free from any larger group or historical narratives – and so, also, free from any group norms that constitute truth and morality.[9]

On the other hand, some have argued that there is no need at all to be concerned about creating a single identity: we should settle instead for a multiple one – a 'protean' self.[10] Proteus was a mythological Greek sea-god who assumed many different shapes in order to avoid having to foretell the future. Proponents of this idea, such as S. Turkle, hold that the constant change and instability of modern life could lead just as validly to a self that is multiplicitous (seen as positive) as to one that is fragmented (seen as negative and medically chronic).[11] Provided there is a way of handling it, there are no problems, they say. This claim is made on the grounds of human resilience, which is greater than one would expect in face of catastrophe and crisis. Turkle also argues that it is entirely acceptable to have multiple selves. There is no unitary self, she suggests: we should think of ourselves as essentially decentred, being 'a multiplicity of parts, fragments, and desiring connections'.[12] Having collected the views of a thousand US citizens in research studies,[13] she argues for the idea of a flexible self and development assisted by online 'play'.

Whether we recognise our experience of self as unitary, plastic, expressive or multiple, our identity or identities are often underdeveloped in one area or another. We might have parents of different temperaments and take after one rather than the other, or we might have been born and raised in a dysfunctional family, or the community where we grew up might have emphasised some aspects of social intercourse over others. Psychotherapy is one way of drawing out those neglected parts of ourselves that we need for well-being and negotiating life. This process sometimes involves role-play or trying out different aspects of identity. There are other ways too that we can experience ways of being that we can't normally, most notably though entering into stories through various media including novels and plays. Children, of course, as they embark on the lifelong process of finding 'Who am I?', quite naturally use fantasy and role-play as means of exploration, and the importance of stories for healthy child development is well researched.[14] In the age of cyberspace, chatrooms and virtual reality offer direct, interactional exploration which is even more attractive as a means of trying out alternative realities. 'Identity play' can be substantially enhanced in imaginary worlds where you can choose your own character-istics and capabilities, unlimited by biological or psychological reality.

Many people report that these 'externalisations of a part of self' have been valuable in helping them, particularly when they are deliberately engaging with problems in reality.[15] One woman described how she came to terms with losing her right leg in an accident, by creating a character in a multi-user domain (MUD) that had lost a leg, then exploring with others what this meant, becoming romantically involved and experien-cing 'virtual love': 'After the accident I made love in the MUD, before I made love again in real life ... I think that the first made the second possible. I began to think of myself as whole again.'[16] Turkle observes that the woman was enabled not to re-imagine herself as whole, but as whole-in-her-incomplete-ness. The play can be used to reflect constructively on everyday life, asking 'What does my behaviour in cyberspace tell me about what I want, who I am, what I may not be getting in the rest of my life?'

Others say that such play helped when, for example, they wanted to be able to assert themselves more or to deal with confrontation. This may be true and likely, as online fantasy and play only elicit certain aspects of self. Some facets of identity are easier to explore online than others. It would be far more diffi-cult to experiment with being relaxed and not verbalising, or doing, while still fully engaging in the medium, for example.

The parts of self externalised in MUDs tend to be the more articulatory and confrontational ones.[17] Case studies on self[18] and research on communication both agree that written telecommunication stresses articulation and assertion.[19]

Identifying people online

A second cautionary point is that online, you don't know who you are meeting. Anonymity, imagined personal backgrounds and alternative identities all hide reality. Although for some people this is an opportunity to work through some personal issues, for others it is a means of concealment for illegal or immoral purposes, such as when a paedophile tries to meet a 13-year old girl by masquerading as one. Even when there is no malice aforethought, the fact remains that the person you meet online may or may not be the same person that you might meet through other media, the phone, video or face to face (as portrayed in the film *You've Got Mail*). In general, the safest working assumption has to be that the 'online other' remains a stranger until you start expanding the content of the dialogue and broadening the number of media used. Nearly all of the high profile love-duets created as a result of initial online contact have moved through the phases of increasing synchronous communication, adding voice and swapping photographs before meeting face to face.[20]

Given that identification of people is difficult online, how do you then decide whether to give people access to information or to do business with them? How do you confirm that they are who they claim to be? To this end, technicians are developing a wide range of identity-checking devices – smart cards, finger-prints, iris scans, and voice prints. There have been suggestions too that our DNA may be used in this way. The purpose of it all is to develop something reliable enough to identify our physical bodies with the electronic persona in cyberspace. This can be done by the possession and subsequent use of a unique object (smart card, digital signature) or by some physical characteristic (eye, voice, fingerprint, DNA). But checking is not the straightforward matter one might suppose.

The managers of such checking devices will use the probability statistics to indicate the unlikelihood of mistakes ever occurring. But they will also be eager to cut costs and so rate the actual chances of, say, two people having the same DNA pattern (as described by the testing equipment using a very limited number of variables) as lower than might actually be expected from the number of people in the world. Technicians are likely to be well aware of the possible error points and the likelihood of their combination – measuring equipment with physical characteristics that lose their sharpness and start to blur data; errors in data transmission to and from databanks; corrupted data in the databanks; and of course the fact that there may be more people in the world than actual or statistical chances, so it could really happen that two people have the same data readings. When two people have access to the same bank account, say, what is likely to happen? There are also problems of inclusion/exclusion. It is possible that a particular test may be inappropriate for a particular group of people – an iris test for the blind; a voice test for the dumb or ill; a DNA test for those who have reason to fear its use more generally.

The problems of 'multiple selves'

Returning then, to the issue of finding identity, Turkle, while advocating the acceptance and development of multiple selves, does acknowledge that the idea may pose some problems. 'Health is when different aspects of the self get to know each other and reflect upon each other,' writes a psychoanalyst[21] – and when individuals have smooth transitions between states of self. We may be 'decentred', but a healthy state is associated with sufficient familiarity with all our parts and to be clearly distinguished from multiple personality disorder (MPD). In developing flexible selves, we need processes for re-collection and for identity creation that makes sense of the various parts of ourselves and integrates them into wholeness. An interesting fact noted by psychologists is that people sleep better when they avoid telephone and other forms of distant communication in the half-hour before going to bed. We don't yet know why this is, but I have a hunch that it may be associated with this idea of re-collection of self at the end of the day. If what we seek is wholeness and integration, then the idea of a Protean self – an identity composed of many selves through which one can cycle and select as appropriate[22] – is not a helpful one. Multiple selves means multiple narratives, and how then can you decide what to do? One of the main uses of a narrative is to understand the present by looking at the past in order to move on into the future. Those I meet in counselling have frequently not clarified their story and so are immobilised by having too many options. Proteus' continual metamorphosis was to avoid foretelling the future: moving on requires wholeness rather than multiplicity.

The second area of difficulty is that in the literature on multiple selves there is a strong reaction against the idea of a moral self, of a person being responsible. Turkle criticises the maintenance of a unitary self on this very ground. She says it produces ' . . . strong pressure on people to take responsibility for their actions and see themselves as unitary actors . . . the pendulum has swung away from that complacent view of the unitary self.'[23] Lyon's 'expressive self', although seeking for an authentic, integrated being, also lacks any connection to a particular moral framework. Moral identities are most often associated with stability and membership of groups with their codes of conduct and norms of acceptable behaviour. In order to get this, some people simply adopt one by joining a fundamentalist group – their identity (to use Lyon's expression) is 'subsumed'.[24] The growth of fundamentalism is identified by Castells as one of the collective identities being formed in response to global capitalism based on information.[25] One of the challenges at this time is how to develop a moral self in the context of a postmodern society.

Finally, at the very time that sociologists have been suggesting some new ideas about how we are attempting to
construct identity – 'plastic self', 'expressive self', 'protean self' –
working against this self-articulation are data-images. These
consist of data about individuals, but rather than the flexible,
adaptable, interpretative constructs that would reflect identities,
they are static, accumulative and tied to the past. The medium
that in part facilitates the new constructs also ties us down into
a history of interactions and logged intentions. Perhaps people
can experiment precisely because they are held by what matters
most in a consumer society – the solidity of financial narrative
against which other narratives are but ornamentation? Data-
images are increasingly important in everyday living, as we
now explore.

Data-image – self as seen
by corporations

In a mediated life, people don't see 'me-myself'[26] – they see my
data-image. This is composed of data collected or acquired
about me and builds up into an 'image' that, in being circulated,
can affect my life-chances and so change what happens in life.[27]
Driving fines in the US affected whether or not people were
accepted to fight in Vietnam; credit-rating databanks can decide
on your chances of obtaining credit, a mortgage or a mobile
phone. In other words, the data-image can present a good
profile or a bad one. The data may be inaccurate, the organisation may apply inappropriate rules which they will not declare, or a software program may have bugs in it – nevertheless,
life-chances are affected by such decisions.

What kinds of personal data are held about me and where
do they come from? Personal data is collected in many areas
of life. Sometimes we give it voluntarily, other times it gets
handed over as part of the transactions of daily life. Here are
a few of the areas in which data gets collected:

- physical aspects such as where I live, the car I own, my family structure;
- my buying patterns (from store cards and credit cards);
- where I go (from mobile phone records, credit card records, car number-plate scanners, and, if overseas, airline records and passport scanners);
- what I am interested in (from URL lists on my computer, clickstream analysis in using the Web, book-buying habits, credit card purchases);
- who my friends and acquaintances are (from communication records – phone and email).

Much of the data gathered by databanks is given voluntarily by the person. Whenever you buy a new 'white goods' product, there is a questionnaire as part of the registration of purchase and warranty form, asking you to help them develop their products better in the future. In the warm glow of a new purchase, many complete details of their TV-watching habits, number of children and their ages, pets, favourite breakfast cereal etc. Many don't realise that this lifestyle data ends up on one of several major databanks to be used by analysts from many different companies and for advertising.

Other data is given in exchange for services or for access to information. Closed-circulation magazines and newsletters, for example, run on advertising revenue and the advertisers pay in order to get to a particular group. The publisher knows that they have that market by selecting people by means of a questionnaire. On the Web, I offer my email address and other details in return for access to information, and this serves as proof for advertising revenue for that website. I even accept that if I want to be able to store online references and bookmarks, or retrace my steps in a large complex database, the program will track my every movement and store it – I hope only temporarily.[28] In these cases I am largely aware that I am exchanging data on what I do in return for access and service. Whether or not this is an equitable exchange is a different matter; here

I simply note that I make choices, some of which are voluntary, which result in data being gathered.

As we go through life, we leave behind a trail of data – from buying goods with credit cards, using telephones, driving through automatic car number-plate analysers and so on. Other data we are forced to supply by law – for example in electoral databases, which in some countries may be sold.

The rules applied to the data tend to be kept secret, but over time some details get leaked. It is now public knowledge that you are considered a bad credit risk if you never borrow and always pay on time – which seems rather counter-intuitive for many people. The more credit you have, the better chance you have of getting more – providing you are paying it back. Less obviously, your physical location heavily influences your ability to get credit: previous bad credit ratings at the same address or in the immediate geographical locality or for members of the same family count against you. This kind of ruling can be quickly overturned but the data and data structure itself is hard to change, despite quite clear inconsistencies. The credit rating companies in the UK have been very slow to acknowledge the need to change them and are only doing so under great pressure.[29]

The rules applied to the data do no more than data matching – they simply compare records of people who fit a similar profile, based on a specified set of data. In other words, they are not targeted at individuals, but at groups or categories of people. Unless your pattern of life conforms to a sufficiently large segmentable group, then you are quite likely to be treated as 'odd' in some way. For example, in the UK, in order to have a TV in the house, you have to purchase a licence from the government and the revenue pays for public sector broadcasting (BBC). If you do not own a licence, you are presumed to be defaulting rather than presumed to be without a TV. This is an efficient assumption, but it casts someone as 'guilty' unless they are able to prove otherwise.[30]

A further development of data-matching is in the use of software that tries to work out what kind of group you might fit into. This is done by searching for patterns – called 'automated inference models'. Whereas this might be OK in marketing, where the only effect on the individual concerned is one of annoyance at being wrongly categorised, there are more serious implications for the development of application to behavioural analysis. In this use of predictive profiling, someone might look at the analysis and conclude, for example, 'this person may commit fraud', because the software tells us that the data is similar to that of a set of people who have committed fraud. This is quite different from the inspection of data in response to some evidence of wrongdoing – for example tax evasion or bogus welfare claims.[31] The latter maintains the assumption of innocence enshrined in UK law and the former, especially if circulated or traded, does harm directly to an individual based on nothing more than predictive probabilities taken from limited data and a software program that is only possible to refine over a very long period. The data-image then circulated is a partial one based on categorisation and prediction. Such profiles might easily exclude pertinent factors not yet identified. Predictive data profiles circulated as data-images are appearing in many places, for example health profiles based on DNA analysis being passed to insurance companies.

My identity is being increasingly tied into a data-image in which I am determined by what I am for others in terms of financial benefits – what I buy, how much I buy, whether I pay back loans, whether I have high lifetime value for the corporation and am likely to generate revenue for it. The terms also include possible financial loss – am I a 'possible fraudster', 'possible criminal', or 'possible person to die young and so you have to pay out money to a spouse'? The driving force behind this is commerce, rather than the well-being of citizens.

At the very beginning of the chapter, we considered the creation of identity in terms of deriving meaning from what we are for others. But how do people find meaning in their data-image? I have three suggestions. The first appropriates the data-image into our identity, the second challenges it with our physical presence, the third demands special protection from speculation.

If we accept data-profiling as part of our new culture, then we could perhaps treat it as we do other parts that require us to operate by a set of rules of sensible practice. In this scenario, we teach children and friends how to construct a 'good' data-image in order to get on in this society: borrow something, pay it back; have a credit card and pay on time. It sounds a little like the construction of a persona for a spy's double life – and it is little different! Graduate students I have taught have been an outstanding source of imaginative ideas for creating false data images – changing names for the supply of utilities to the home (this was possible in the UK until recently); always giving false demographic data (age, sex, likes/dislikes); making up the names of a company you belong to when going to conferences (this enables you to track junk mail over years); changing email addresses on a regular basis (necessary when in chatrooms). All these (the only ones which I am prepared to publish) are useful in examining how data-images are created and passed around and how they come to be collated.

In addition, we could also create an ethos of allowing appeals (with the person present) as a matter of course, rather than it being treated as exceptional. Some women have found, for instance, that when separating from husbands, their data record is either non-existent or tarnished with problems, so that they find it hard to borrow money at the very time they need help most.

Finally, predictive data-images, unrelated to the reality, should have cautions attached to them to indicate their predictive status (on screen, especially, it can be impossible to tell).

Boundaries of self are not defended by privacy

Issues of identity are closely related to the projection of self across the private/public boundary – in other words, self-disclosure. What we disclose, and when, relates closely to our identity. Data can be advanced, or delayed, or maybe concealed altogether – as a part of a transmission process. If data collected from the normal course of living is transmitted by someone other than the individual concerned, then the individual is not acting in self-disclosure, but being disclosed against. A letter that arrived at one home from a country hotel with a special offer to a regular customer had the wife on the phone to her solicitor filing for divorce. The difficulty here was that there could have been other explanations than the wife probably assumed, but the discussion between spouses was changed for ever by the order in which data was disclosed.

The transmission of data in this way raises again the issue of privacy. A database on Computer Ethics has over 1,200 items (and does not include most of the European writing on the subject). Within this collection, the largest categories are privacy and the autonomy of the individual. We have already briefly considered privacy from the viewpoint of attention (in Chapter 4); I now want to consider it from the viewpoint of self.

Privacy has not turned out to be a very useful concept, despite the quantity of writing on the topic. First, privacy is not intrinsically a good thing. Absolute privacy – solitary confinement – remains one of the ultimate punishments. It therefore seems an odd concept upon which to build too much – especially in the social context of widespread concern over the need for more communal life, for sharing, and for transparency. Nor has the concept of privacy been used to make any useful distinction between the private life and the public life of individuals.

The blurring of the boundaries between public and private life is highlighted by questions of identity and disclosure: do we include our data-image as part of our identity, even though we have no direct control over its disclosure, or do we exclude it and come to terms with a 'leaky' identity?[32] Where does the boundary of myself lie? Am I responsible for my 'good name'? If so, then we need to find ways to lay claim to the cyberspace equivalent – our data-image.

On the other hand, there are clearly false or projected images of myself, from which I would want to distance myself: the predictive images that may alter life-chances, and others (involving wrong assumptions based on statistical analysis) that deny citizens the freedom to go about their lawful business without harassment.

The evident erosion of the private/public boundary gives the lie to the idea that we can be autonomous. We might be seeing an end of the 'individual humanism experiment'.[33] Use of systems and networking subverts the concept of separate individuals – so much so that one writer sees the interweaving of online interactions with others as leading to a 'saturated self': a self that contains parts of many other people.[34] Even as global capitalism seeks to enhance individualism in consumers, there is a search for new connectedness around shared constructed identity. If this is so, might not the current emphasis in computer ethics on 'individual' actually be part of reviewing what it means to be an interconnected self, a reaction against a change in the concept of the autonomous self?

Influences on the concepts of identity and the nature of self now come from many different directions: change in work, virtual therapy, data-images, consumerism, interconnectedness, and the search for collective identities. Such changes are bound to affect us because we can no longer think of an ego that independently sets itself in a place; identity is formed through interaction and exploration. We know who we are only in relation to others and how we think about those relations. We have 'a self that recognises itself only through the multiple ways the other affects the understanding of the self'.[35] That 'other' includes other people, those we met online, our past history, including our data-image, and the speculative discourse of possibility in our own selves, such as is explored online in chatrooms and MUDs. Rethinking our identity could be problematic if there is no reason to choose one direction over than another.

To say that we are historical beings who determine our selfhood as we go along, is not to plunge us into ethical relativism, for in Christianity there is, if we may adapt Macquarrie's words, a 'constancy of direction'. The direction may be grasped through various categories of thought (apocalyptic, ontological, existentialism), but the goal remains constant.[36]

We explore this further in Chapter 8.

The self we understand in relation to others is always an ethical one – we can turn away or towards others, we can be for or against them (and the 'others' can also be taken to mean parts of ourselves).[37] When ethical norms are not well established, as they are not in the area of IT and cyberspace, then an identity which has the self-esteem that comes of knowing in what ways it helps others is able to respond on behalf of others, and then move towards demand for the right treatment of others. This takes us on to the subject of justice and how we participate in politics related to information technology.

chapter seven

Political Engagement

How does information technology affect our political participation? We need to find ways of governing cyberspace to
avoid domination by corporate business, and also to avoid
use of technology by governments to control citizens, so how
we participate in the future of cyberspace is important.

What can we do to express 'democracy' through electronic
communication? There are advantages as well as disadvantages in the use of cyberspace in this.

Ordinary people (i.e. not the technical elite):

- can gain 'situated knowledge' as users of particular
technologies;
- can use information technology to create or participate in
'networks of concern';
- can and should engage the technical elite in various ways
to develop IT for the benefit of all.

'THE ROLE OF THE nation-state will change dramatically and
there will no more be room for nationalism than there is
for smallpox,' wrote Negroponte.[1] He was commenting on the
perceived trend of globalisation, and the changes he refers to
are in the geographical character and the boundary-marking
that make up states.[2] And yet there is much evidence that points
in the other direction. Many of the conflicts in the world are
about nationalism – over who is a member of a society and
who is not, and over the borders that divide different groups
of people. A people's identity and sense of belonging are not

being absorbed into a new 'internationhood' but, in opposition
to globalisation, are becoming more marked.[3] So what is going
on? Castells argues that the globalisation of capitalism based on
information technology and the search for identity actually
belong together as the two main marks of our time.[4]

Many states find themselves among new players in the global
field as a result of the globalisation of corporations and trans-
national movements such as environmentalism and feminism.
To enable the country to flourish, states have to participate
strategically in the global flow of capital, and trust that the
improved economic environment benefits everyone. This
trickle-down approach takes time, and delayed gratification
through corporate success appears to citizens not to represent
their interests directly. It can make them react and want a clearer
identity. (One could, for example, see the debate in the UK
about Europe in this light.) On the other hand, if a state con-
centrates on its identity, it can fail to engage with and maintain
competitive advantage in the planetary economic scene.[5] To
flourish, states must be more than an expression of the collective
identity of a people by means of boundary-marking. Even
though states are not the only players on the international
scene, they are not powerless either, as the pundits on the new
cyberspace would have us believe. They still act as nodes of
economic power with other states and international bodies.
Economically a state can still do many things; for example, it
can ease, slow or block movement of capital, labour, information and commodities using its regulatory powers such as
those that provide an educated workforce and an environment
conducive to business. Economic considerations are just one of
the duties of a state; others include military security (often the
dominant role in people's minds), political, cultural and
environmental concerns.

> In addition the state is subject to other, less definable in-
> fluences: networks of capital, production, communication,
> crime, international institutions, supranational military
> apparatuses, non-governmental organizations, trans-

national religions, and public opinion movements. And below this level, there are the influences of communities, tribes, localities, cults and gangs.[6]

The idea of a state is perhaps changing towards its becoming a node in a broader network of power, but there is no sign of a dramatic change in the direction of its decline or even death.

Welfare and protection in cyberspace

One reason people think that the nation-state may be decaying is that it appears to lack control over the Internet and what happens there. 'There can be no laws about cyberspace, because no country can control it; it lies outside the jurisdiction of any single country or organization.' Thus runs the argument for the futility of efforts to exercise legal and political control. Any attempt to introduce legislation would result in rapid movement of Net nodes to more benign political and economic environments. While working for an international publisher, I wrote a letter on their behalf to the UK government, saying that if they were to tax electronic publications by VAT (value-added tax), we would move our operations to the USA, where there was the promise of at least a two-year moratorium on all taxation on the Internet. The attempt to tax is futile, I argued. I also stated that I would like to be part of a country that was encouraging the development of electronic services, rather than penalising them with tax that was not applied to books and other publications in the UK.

Nevertheless, as a citizen I do want government to ensure that corporations and commerce in general cannot just turn the Internet off, in the same way that electrical companies cannot turn electricity off to suit their purposes. I want government to control the power of key cyberspace organisations, in the same way as they do media and other technical companies. I want my personal data to be protected by the use of law if necessary. In short, I want government to care about my well-being with sufficient authority over cyberpractice in order to do so. I want to know too that my use of cyberspace, and the parts of my life where it is integrated into daily living, will actually be given the conditions in which to thrive, just as well-being in the physical sphere is safeguarded by conditions set up by government regulations (for example, through safety and health protection and services to help heal one when ill).

Equally I want government not to exercise undue force or control on me or other citizens; I want to be free to pursue all that is legal without being watched and studied; I want to live without fear of making a mistake and to be able to express opinions and views that contravene no laws. In short, I want to be certain that I am protected by the rule of law from the excessive use of power by governments.

Finally, I want to be able to participate in choosing the kind of government we have and the laws that are made and the decisions made about technologies that shape my daily living.

These three areas I have just described reflect the three-way relationship that individuals have with the state. In this individuals are:

- creators of state authority through the democratic voice;
- potentially threatened by state force or coercion, but protected by rule of law;
- dependent on the services and provision for welfare organised by the state.[7]

In other words, I want to be part of a state which I help create, which provides services to me, and from whose abuses of power I am protected. This is a long way away from the views of the libertarians of cyberspace who believe in total 'freedom' and the absence of any governance. The freedom I want is both protection from the abusive power of others and space in which to develop an abundant life.[8] We therefore need to find ways of governing cyberspace as part of living. The two main fears implicit in my wants are that technological development will be dominated by corporations with the support of governments in a continuation of scientific management and/or that government itself will choose to employ technology to exercise control over its citizens.[9] In the UK, the Regulation of Investigatory Powers Act 2000, with its demand for anyone to hand over encryption keys if it is so requested, and with prison sentences for failure to do so, was greeted by most with dismay, as excessive use of power. It was developed not by the department dedicated to promoting business or welfare, but by the one responsible for law and order and the secret service, and was ushered though with little public discussion.

All states demand the right to monitor communication activities if they have a suspicion that citizens or aliens may be undermining law and order. In order to do so under the legal systems of the UK, application must be made to obtain a warrant, and records kept of that transaction so that the power of government does not extend beyond its remit under law. This means that if a telephone line is to be monitored or letters opened, a formal request has to be made for the right to do so. The Regulation of Investigatory Powers Act 2000 treats voice and data in exactly the same way. However, the technical underpinning of the Internet standard does not necessarily allow the little bits of each individual's message to be separately identified, so it is likely that Internet Service Providers will fit a 'black box' to collect all email traffic and supply it on request. Although there are legal restrictions on the analysis of the content of messages, no such limits apply to the 'click-streams' that trace someone's movements around the Web. This means that the UK government can theoretically follow the line of movement and so of the thought of users. Moreover, encryption keys used for ensuring that there is no tampering or sight of messages may be also demanded upon request; Ireland, however, has passed a law saying that the government will make no requests for encryption keys.

All of cyberspace emanates ultimately from code. Technical decisions by programmers about code actually change our experience of daily life. Understanding how to participate in these decisions is the essential task facing governance of cyberspace. What kind of space can we create? Cyberspace is enacted in hardware and software code and enables or disables future courses of action.

> This code presents the greatest threat to liberal or libertarian ideas as well as their greatest promise. We can build, or architect, or code cyberspace to protect values that we believe are fundamental or we can build, or architect or code cyberspace to allow those values to disappear. There is no middle ground. There is no choice that does not include some kind of *building*. Code is never found; it is only ever made and only every made by use. As Mark Stefik puts it, 'Different versions of cyberspace support different kinds of dreams. We choose wisely or not.'[10] [11]

The remainder of the chapter concerns the processes by which we can choose wisely. I will consider the use of IT and cyberspace in the political process and then, and more importantly, address ways in which we can participate in the design and development of technology. Since technology is changing social space, one might expect a public discussion. Historically, however,

> ordinary citizens have been excluded from key choices about the design and development of new technologies, including information systems. Industrial leaders still indulge the old habit of presenting as *faits accomplis* what otherwise might have been choices for diverse public imaginings, investigations and debates.[12]

In what ways can we participate in the future of cyberspace?

Moving democracy into cyberspace?

Electronic referenda, flash-polls, the people's democracy: these are all expressions that signify the idea that the people, not their representatives, are now able to take political decisions through the use of electronic communications.[13] In its more radical version, this view argues for cyber-voting on every decision. This is one possible form of democracy, but there are many others – as can be seen by the wide variety of political structures within what we call democratic countries. Although the act of voting could be readily transferred to electronic means (assuming that there is no exclusion because of lack of technology) there are other aspects to the political decision-making process that cannot be transferred so easily.

On the face of it, cyberspace looks like an attractive area in which to develop the democratic ideal of the commons – it affords access to information in order to understand the evidence, exposure to ideas you would not normally choose and a means of weighing up the likely impact of any decision on other legislation and decisions. But each of these three has its difficulties.

First, information certainly can be and is put online by many democratic governments. In this form, however, it may not always be as accessible as we might suppose. We can easily be impressed and overwhelmed by the quantity as well as get distracted and fail to notice what is not there. If we cannot find something it is often easier to believe it is we who have failed, rather than the supplier of the information. There are also many tricks that can be played – putting information onto the Web without links, so you can access it only if you know the exact page address, or putting it into an relatively obscure area. I am not suggesting this is done routinely, but I mention it to show that the ideal of ready access to information can be thwarted both by our own actions and by those of the provider – in each case we will not necessarily realise.

Second, cyberspace is set up to provide self-selection of information. Will people actually choose to seek out and consider the contrary views of others or (more probably) will they seek out those that match their own and so support them with greater force? There is no reason to expect that accessible information will act as a check on irrational public opinion or behaviour in a democracy – indeed fragmentation even to the point of moral anarchy could be the result.[14] Segregation by self-selection already occurs in US television: of the top ten TV shows watched by white people, and the top ten watched by black people, only two shows are on both lists.[15] Most commentators believe that the asynchrony and self-selective aspects of new technologies mean that we tend to avoid exposure to the uncomfortable and challenging views of others. An important aspect of the democratic commons is the opportunity to have that exposure and exchange views on matters together, as in Athenian-type discussions (albeit not predicated any more on slaves doing all the work, which gave the Athenian male citizens time to engage in politics). Perhaps if we could find ways to use cyberspace for direct discussion it would help to avert the threat of fragmentation? The evidence suggests not.

Third, deliberative democracy is proposed as a way of engaging online in debate and decision-making. However, the medium of cyberspace, as we saw in Chapter 5, is far better suited to presenting information and opinion within divergent thinking, than to moving towards convergent, mutually agreed decisions. In order to move towards resolution, discussion has to be structured for convergence and this quickly makes people feel that alternatives are not being fully considered ('why this proposal and not that one?'). Moreover, in the opaque and mediated world of cyberspace, it is easy to believe that other speakers/writers are given a preferential voice/write or that there is some computer software that disadvantages one's own attempts (easy enough to build). The process is anything but transparent and unlikely to be successful without the trust that comes from total transparency.[16]

Thus, what evidence we can muster suggests that cyberspace does not provide any return to the ideal of the commons. Problems with access to information, exposure to challenging ideas and online debate, all dash any hopes that we could return, in some sense, to a more popular form of political decision-making. One suggestion to compensate for the deficiencies in cyberspace is to extend the idea of the jury and take a sample of citizens away for a long weekend of deliberation, where information can be presented and provided on request, and discussion take place face to face, with all the emotional force of presence. In this way, they would be equipped with all they needed to move towards a decision.[17] Only in such a way, it is claimed, could the media's urge towards polarisation and oppositionalism be converted into exploration and consensus.

Turning to the representational aspects of politics, we might consider the advantages of easy communication with those we elect to take decisions on our behalf. Email, in particular, is an easy way to write and make our views known to politicians. It also makes it easy for them to avoid reading. US congressional staff routinely block email from all except those who can vote them back into office – their constituents. The content of the email is then subjected to an automatic content analysis, by use of a software program, which identifies the writer's issue, adds it into a tally of such issues and then replies with a (hopefully) appropriate letter.[18] So the email is read, but only by a machine, and received only in the sense that a digit is added to voter opinion on subjects predetermined by the programmer or some categorisation scheme.

Political parties explicitly court the views of voters in order to improve their political marketing. They can check how a message or policy will be received, by strategic polling of a sample set of voters. By adjusting the message and polling different samples in a continuous cycle of feedback, they engage the voters in helping the party to make the best impression overall – and thus to feel part of the political process.[19] The optimisation of the message is essential to the party's future in the competition for credibility among the voters. It is argued, however, that the media, including new Internet technologies, 'frame' the political process.

So are there any features of cyberspace that could suit and serve the political process? Internet suits short-term attention and quick responses. In this it is ideally matched to a *monitorial* approach to political participation, in which a relatively small group of people, aroused by some high-profile situation, mobilise tens of thousands online. This works best in areas where citizens feel comfortable taking positions without further information or deliberation.[20] These might be overtly political, as in the successful mobilisation of US citizens over Clinton's impeachment and, as I write, the move towards reform of the US election system in order that the people's vote, rather than the electoral college, should elect the next President. Or they may be non-political, as in more general Third World or environmental causes, for example Jubilee 2000.[21] This approach is one of the ways in which Castells sees the political use of cyberspace.

He points to three directions where there is a natural fit between politics and cyberspace: political mobilisation around 'non-political' causes; the enhancement of political participation by horizontal communication; and the re-creation of local states.[22] We have suggested that short-term political causes appear to be effective on Internet and I suspect that this is also the way that horizontal participation may develop – that is, not in any democratic style of deliberation by the people as a whole, but in dissident or marginalised groups finding enough consolidation to make their opinions known. This, too, is the route that I believe is most effective for public participation in shaping the future of technology itself.

Mechanisms for the shaping of technological society

The masters of technical systems have far more control over the design of transportation systems and the selection of innovations and so of our experience as employees, patients, travellers and consumers than any electoral institutions.[23] Many of the choices made will not even be explicitly considered – decisions may be taken automatically as the next step in technical development, or as what it is considered the market will bear. Nevertheless, choices are made every day about the relationship between people and new technology, and those relationships shape the kind of society we will live in.[24] On reading that the telephone companies in the USA had agreed that fiber co-ax was the right design for their services, Winner acidly remarked, 'How reassuring; evidently the "right design" is headed our way again and we have not had to lift a finger.'[25]

He argues that the populace should not settle for an effrontery as blatant as to exclude them from technological decision-making, especially for the spurious reason that the technical elite are the only ones who have enough knowledge to make the decisions on everyone else's behalf. Technological decisions go far beyond questions of efficiency – they shape our social environment and the life patterns of citizens.

The result of expertise legitimatising power in society is that organisations are constructed and reconstructed around the paradigm of technical administration.[26] That is to say, the 'efficiency' paradigm of scientific management is passed from the factory, with its division of labour into managers and the managed, into society, with its technical experts and those expected to participate without comment. The outcome is one-way communication and a lack of opportunity for discourse and debate about the kind of society people want.[27]

How might the non-technical get involved in technological decision-making? Some already do and they provide a model. For example, people form themselves into groups in order to protest about the course of a new road and suggest alternatives. Looser groupings submit to a website all the bugs they find in certain bits of software or of an organisation's handling of data, thus revealing patterns of error in design or decision-making. These groups coalesce around a particular concern and so are called networks of concern.[28]

In democratic terms, this kind of participation sits uneasily with representation, which is defined through local geographies. Networks of concern cross the boundaries of local geographical regions, which remain the primary unit for election of representatives (although not the only one, as seen in the UK elections for Members of The European Parliament, where the country's overall voting is taken into account in order to give an overall as well as a local party representation). This is a reason why representative politicians have found it hard to address the technological issues raised by networks of concern; they may not feel authorised to do so unless the focus of the concern happens to be in their constituency.

One helpful way of examining the contrast is to consider Feenberg's suggestion of comparing the spatial boundaries of representation with the temporal historicity of technical special-isation.[29] In the former, since states are large, geographical units are created to be small enough to engage the interest of local citizens and animate discussion. In the latter, technical develop-ment is vast because of the accumulated knowledge and expertise necessary to understand large and complex systems. Those take time and so there is no possibility of a lay person directly gaining the required technological expertise. They can, however, as users, gain 'situated knowledge' of particular tech-nologies. For example, they might become expert on the effects of a particular drug, or a software program, or the proposed solution for transport bottlenecks, or public access to Internet. Feenberg argues that the global/local division of representation can be reflected in a technical historical/situated knowledge division for technological participation. In the same way that democratic representation has found ways to handle the local in the context of the total citizenship, so too there could be ways of handling the temporarily situated knowledge of con-sumers and users in the context of the expertise of technicians. This resonates well with use of the Internet to obtain in-formation to support situated knowledge and communications to create a network of concern.

Examples of successful change effected through networks of concern have included the transformation of data networks into communication networks, barrier-free access for disabled, access to experimental treatment for AIDS sufferers and, in the UK, availability of food free from genetically modified organ-isms. Key shifts in perception were needed to translate issues from being personal or private concerns into wider ones about the quality of life sought by citizens, and from being technical decisions into technological ones that are legitimately discussed by lay people. How are those moves accomplished? First by creating networks of concern and developing situated knowl-edge, and second by tactics that engage the technical elite.

Tactical involvement in shaping cyberspace

There are a number of tactics for seeking closer involvement with technological development, including provoking technical controversies such as those over environmental issues, having innovative dialogues with designers, and making creative appropriations of technologies in some way. We briefly consider each in turn.

Technical controversies have often been initiated in areas in which citizens feel themselves to be medical victims or potential victims of some technical process – which might be chemical or nuclear, or engineering (such as the position of gasoline tanks in cars in USA, brought to public attention by Ralph Nader). The controversy is set between the health interests of people and the autonomy of technical organisations:

> Normally only technical professionals would pay attention to industrial processes environmentalists challenge, but today we believe in the right of the public to prevent such processes from doing harm. To be a citizen is to be a *potential* victim. This is why information plays such a critical role in environmental politics: the key struggles are often decided in the communicative realm by making private information public, revealing secrets, introducing controversy into supposedly neutral scientific fields and so on. Once corporations and government agencies are forced to operate under public scrutiny, it becomes much more difficult to support dangerous technologies such as nuclear power.[30] (Italics in the original)

In the last major public enquiry into nuclear power in the UK (Sizewell B), there was often surprising agreement between environmental groups and the nuclear industry over facts, but their assumptions, about future profitability, cost of disposal of spent fuel and decommissioning, and safety, were the decisive differences. The nuclear industry was forced to declare the assumptions they were making. Technical controversies encourage the release of information so that lay people can participate in debates.

In considering innovative dialogue and participative design, it would be easy to imagine all technical staff as another breed, separate and distinct from users and from the life-world in which new technology will be introduced. This is far from the case: technical people do engage with the public on issues they wish to raise. For example, when the scientific paper announced the existence of an adult sheep called Dolly, who was the result of a successful cloning experiment, the lead scientist was already a member of a group discussing the ethical considerations.[31] There are examples also, of experts who have turned against the whole of their endeavour or even blown the whistle on the impacts of their organisations. These include the scientists who developed the atom bomb and then went on to create and motivate a network of concern, and the computer scientist Weizenbaum who delivered the wake-up call in computing and sought to engage others in the choices being made by programmers.[32]

Of particular note are those designers who seek to include the users in the design process itself; this is called participative design. This is often unpopular with management, because the potential market learns how things are done and also engages directly with technical staff instead of through marketing and sales representatives who are taught to say the 'right things'. There are other difficulties of course: as every software designer will have experienced, what users want changes the minute they get a chance to use what has been delivered to them. Then the designer complains that the users are inarticulate and do not really know what they want, and the users reply that they cannot know how it will all work and fit in with their life-world without trying it. But in the users' appropriation of the technology, new possibilities and requirements come to light. Constant engagement with users, therefore, is a means of generating useful new ideas and ensuring that the service does at least some of what the customers or users require.[33]

Finally we turn to appropriating technologies. I have already discussed the taking over of data networks for communication, and one can see small, sometimes very unexpected appropriations in many places.[34] One example is the way that school students used SMS-text-enabled mobile phones to pass notes around in class – between a teacher asking a question and receiving a reply, the answer might have been passed to the respondent by phone (yes, it has occurred). The spreadsheet on a PC for accounts was an unexpected appropriation that led to a different kind of technical development than was originally envisaged for software products. Instead of only completed software customised for each organisation, it opened up the possibility of a generic one with basic capability. Access to experimental treatments by AIDS patients is another major example of appropriating technology, given by Feenberg.[35] When AIDS first emerged as a problem in USA, the medical profession was uncertain how to meet the need to care for those it could not yet cure, especially when they were demanding experimental drugs. AIDS patients requested access to clinical research and were prepared to forgo the normal requirements of minimum exposures to new drugs, statistical balancing with patients taking placebos, and use of subjects with no prior history of participating in experiments. AIDS patients already had strong networks of concern built up around issues of gay rights, and they were able to mobilise these so that the FDA (Federal Drug Administration) opened dialogue with them and AIDS activists were placed on major committees. Technical practices in the medical profession were then altered to allow these activists access to experimental treatment.

In summary, there are real possibilities for people to engage with technological futures. We know they are not hypothetical possibilities because we can point to instances where this has happened. To do this means setting up networks of concern focused around a technology (or one aspect of it) and developing situated knowledge. It also requires some processes for engaging with the technical elite, either by provoking controversies so there is public attention, or by appropriating the technology in such a way that the direction of the technology is changed, or by working together with designers in participative design and innovative dialogue.

Technical leadership has a distinct and essential place in the division of labour and cannot be replaced by popular action.[36] The intent of this discussion is not meant to blacken or accuse people working in technical areas, but to find a way of allowing the public to participate in the decision-making. The excesses of autonomy can be reduced so that there is a tactical involvement by customers, consumers, users and all those affected by the shaping of the social environment in which they live.

Cyberspace is better used indirectly in the service of new structures to involve people in their technological futures, rather than directly in the established patterns of democratic procedures. Instead of effort being directed towards cyber-voting and the like, it would be better expended establishing routes for the people at large to discourse with experts on the futures they want. Cyberspace can be used to create and sustain networks of concern – an unusual opportunity (*kairos*) where the need and means fall together:

> [I]t is evident that, for better or for worse, the future of computing and the future of human relations – indeed, of human being itself – are now thoroughly intertwined. Foremost among the obligations this situation presents is the need to seek alternatives, social policies that might undo the dreary legacy of modernism: pervasive systems of one-way communication, pre-emption of democratic social choice [by] corporate manipulation, and the presentation of sweeping changes in living conditions as something justified by a univocal, irresistible 'progress'. True, the habits of technological somnambulism cultivated over many decades will not be easily overcome. But as waves of overhyped innovation confront increasingly obvious signs of social disorder, opportunities for lively conversation sometimes fall into our laps.[37]

chapter eight

Extra-connected Living

Cyberspace is not finalised: this is our opportunity. Important choices remain: this is our challenge and our reponsibility. In what direction should we move? There are some false directions ('temptations'):

- not seeing the future as open;
- thinking we can solve problems as though we were the technical experts;
- thinking we can grab a simple 'what-to-do' framework.

For those pondering on the best direction to aim their activity in relation to cyberspace, what are the implications of being 'connected to Christ'? As members of the community of faith (church), Christians have a place to work things out with others, and to learn how to talk about what is happening. As individuals in their personal beings in cyberspace, they have in Jesus an example of the kind of non-controlling, invitational power that needs to be sought and practised. The task is possible, and worthwhile – and love does indeed come into it.

MOST KIDS WHEN THEY enter a big space will break into a run or a dance. The space has a form and it says to them that it can accommodate their activity; it calls forth a particular response. Beneath a Gothic arch, people are drawn into standing tall, while gazing upwards and feeling small. In a forest glade – a room enclosed by walls of trees with an open roof of sky – we feel invited to linger. Other spaces elicit a more negative response: in subways or narrow passages between tall buildings, for example, we may feel like hurrying in fear or apprehension. And of course different people can have differing responses to the same type of space. Cyberspace too has a space-form to which we respond, and it is one with many facets. It is a space to explore – we click and browse; we find answers to questions and enjoy the anticipation of serendipitous discovery. It is a space in which to share information in words, pictures and sound – read that, listen to this, click there, look here! It is a space that accommodates all the activity of a market – buying and selling; window-shopping and ordering. It is space in which to play – pretending to be a fantastic mythical creature with magical powers, or engaging in international team games. Predominantly, it is a space that says 'communicate' – feel free to talk to strangers. It is a space that has incontrovertibly altered the social landscape, even at this early stage of its development.

Different spaces call forth different types of activity; they also emphasise certain aspects of person. Little kids in big spaces physically feel the joy of their bodies and under gothic vaults people become conscious of their historical smallness. In cyberspace, bodies are superfluous and as we are encouraged to leave them behind, we experience it as a place for the conscious articulating mind. The form is symbolic (words and pictures describing things which are not actually present) rather than sensual. The form of cyberspace also presents as an all-knowing silence – storing where we have been, what we have done, who our friends are – and as an endless demand – grabbing our attention, interrupting life in the material world – through phones, mobiles, WAP devices, advertising, government legislation, and the demands to keep changing equipment because some standard has altered. As a result we may also experience it as intrusive, rather as an over-zealous policeman anxious to catch us out, or perhaps a needy child wanting our attention.

These aspects of the new space are the forms we noted in the lineages in Chapter 2: the Internet that says 'come play'; urging escape from the 'here and now', leaving bodies behind;

and the continuation of scientific management that requires monitoring and control of consumers and citizens. Some will observe, quite rightly, that there are certain areas of cyberspace (such as full immersive virtual reality) that do engage the body fully, but these are not yet subsumed within the social spaces, nor yet are they expressed as strongly in a developmental lineage with attendant utopian hopes.

There are many approaches to looking at cyberspace, but the clear impact on social spaces is the one that I have adopted as being particularly useful and that underlies the main themes of this book: of seeking knowledge through lived experience that affirms the importance of bodies; of eschewing controlling power to stay in a position of invitational power; and of being involved in shaping the future by tactically engaging with the powers-that-be.

With this new space come new questions to answer and choices to be made: what is appropriate behaviour? who owns digital content? who are the people to whom I attend? how do I express who I am in this new space? who looks after it all? who will make sure it is there when I want to visit? who will protect me from harm? These are decisions in which everyone in society has a stake; their results will shape the daily experience of every man, woman and child. I have touched on some of the issues in the chapters of this book, but even to catalogue, let alone discuss the decisions to be made with respect to cyberspace is far beyond the scope of this, or any single volume. One could, for example, discuss the questions of cyberspace for data-driven organisations – particularly for scientific and medical research and development (such as genetic engineering and new drugs); for automated control devices (for example the Thames Flood Barrier, or body implants that release drugs into the body when triggered by events such as rising blood pressure – which could be used in order to control anger-based violence); for finance or internal business operations; for military practice and security; for the government's increasing dependence on and use of IT; and for entertainment and media. Rather than be drawn into any single area, I have instead concentrated on the social spaces directly affected by cyberspace and have attempted to show how and where the options requiring decisions emerge. Thus, of necessity I have been highly selective and I do, therefore, recommend use of the annotated bibliography and exploration of other books.

One of the primary aims of this book has been to demonstrate that cyberspace is not finalised: its form and function are yet to be fully determined and the norms of behaviour in a society using IT have yet to be established. That is the opportunity we have. But it is also an appeal not to leave the decision-making to those who happen to know about the technical aspects, or those who have the power to ensure that they are the ones who benefit. That is the responsibility and the challenge. Such an appeal would be useless without an assurance and an explanation of how people unversed in technical detail can participate in deciding how to code cyberspace. Thus, the underlying argument is that there is a position to be adopted that can effect change – both through everyday living and by becoming involved in political activity. That position is one of engagement through creating networks of concern, gaining technical knowledge situated in that concern, and then finding ways to interact tactically with decision-makers.

Assuming then that such a position can be secured, the question that follows is: in which direction should the influence be exerted? Should it be, for example, to seek for everyone to be on the Internet so that they can have the benefit of participating in cyberspace? Should it be to reject such artificial new forms of society and to return to a revised form of community-based living? Should it be to make assuming a false identity a prisonable offence? Should it be to use IT to monitor every citizen so that criminals are nearly always apprehended? Should it be to use IT to ensure that everyone in society spends enough so that the society flourishes economically? In other words, what kind of ethical basis should we adopt for making these kinds of decisions? Perhaps, given the underlying themes and messages of the book so far, it may come as no surprise that I am not going to offer any easy formulas or a 'one-size-fits-all' ethical framework in order to answer that question. Even if one were within our grasp, I believe that the desire for a strong and clearly defined ethical framework is the equivalent of the desire to be in a dominant position of controlling power within the technical domain.[1] What I do suggest, however, is that commitment to the position of non-controlling and invitational power, with its requirements of relating and personal responsibility, is itself a strong enough paradigm from which to advance many of the decisions.

In this final chapter we explore the themes more explicitly from a Christian perspective, and consider Jesus as archetype of the subordinate position of power, as both justification and model for this approach. We then look at how this commitment to non-controlling and invitational power can be worked out, first through communities of faith ('church') and then through individuals in their personal being-in-cyberspace. The more usual order is perhaps to begin with the individual before moving on to the group but, in this instance, it is within the communities of faith that the tasks of working out morality and modelling patterns of living are done – and then individuals and organisations can participate with others in the political processes used to establish societal norms. Before this, it will be helpful to recognise some of the pitfalls in examining ethics in a technological area.

Resisting temptation

There is a collective responsibility for our future, inherent in our social and cultural being; we share together obligations for cultural vitality.[2] As we begin to take on that responsibility in the area of technology, we are vulnerable to particular temptations: first, not seeing that the future is open to shaping; second, leaping directly from encounter with cyberspace as user into solving its problems in the assumed role of technical elite, and third, thinking that we can grab an ethical framework that will tell us what to do.

To begin with the future: there is none except that which we choose to make. That choice is not completely open or free: it is in part determined by decisions that were made in the past; we are only able to explore and express certain possibilities. Nevertheless there are many possible futures because IT and cyberspace, like all technologies, are socially constructed. Luddite has become a pejorative name, being seen as (in the title of a book) *Rebels against the Future*.[3] However, as the book makes clear, the Luddites were in fact rebels against *a* future. In 1811, people were cut off from the commons, so unable to forage for food and to support themselves as self-employed workers. Because of high food prices they were then forced into factory employment with poor pay and appalling conditions. This was the future which the Luddites were protesting against: their appeals were to the government of the day to help them in their distress and their actions were to make the government take notice of their suffering.[4] There were alternative futures, for example one that included financial support for the starving. The government, however, did not heed the appeal and so forced labour into factories – a critical decision that led to corporations having the status they enjoy today. It is also the reason why General Ludd's name is remembered when many other protesters have long been forgotten. Neo-Luddites are likewise accused of being against technology *per se*, but they too are protesting about a view of 'the future' which is regarded as not open to discussion and which is presented as if already set in stone (or immutable code). But indeed it is not. Despite the hype and the advertisements, there are many examples, some of which I have described, of ways in which the future was not predetermined in the past, and therefore it is not now.

Second, in beginning to discuss cyberspace and becoming aware of the problems, it is very easy to jump from encounter with the technology as a user into solving the problems from a strategic view. It is hard to resist imagining ways of controlling whatever pain or frustration is fresh in the experience. But modern technologies have two sides: one the strategic viewpoint of the industrial leaders and managers of the systems, and the other the tactical support of the human beings they enrol, or anticipate enrolling, as users, customers and consumers. The first understands itself in 'objective terms as knowledge and power; the second yields its secrets to a phenomenology of lived experience', writes Feenberg.[5] It is a continual challenge to remain in the second, living and acting from below in relations of power; staying with the experience, rather than leaping in to 'solve' problems.

There are many books that perceptively describe encounters with cyberspace from the viewpoint of user, and then move directly to the position of self-appointed master-controller of all knowledge and power in order to redesign the system on behalf of everyone else. This approach rarely solves any problems simply because the individual, however knowledgeable or expert, is no more representative of the wide range of human experience than the original designer.[6] The 'solution' still does not take account of the wide variety of people's thoughts and beliefs and their frequently conflicting views and uses of systems. Moreover, to assume such a position is often to misunderstand the technology itself. In my involvement with people working on technical systems, one of the things I do most frequently is to encourage them to maintain their position as users: I urge them to articulate what it is they are trying to do instead of guessing at technical solutions. Moving away from the position of user also perpetuates a technistic approach – identifying problems as things to be fixed with further technical solutions. The attempt, for example, to control access to pornography on the Internet was initially framed in exactly these terms – as a 'problem' which required a technical 'fix'. Several fixes were implemented, but none very satisfactorily, and take-up has not been as great as expected, largely because of further problems created. Moving away from the experience and into the strategic position of power can therefore exacerbate existing problems and introduce new ones.

Third, just as it is hard to resist jumping into technical 'fixes', so I suggest there is an equivalent temptation to reach straight for 'easy' ethical solutions. It is as easy to see pornography, for example, as an ethical problem to be fixed by ethics, as it is for us to see it as a technical problem to be fixed by technology. But the reality is that there are no easy normative ethics to get hold of. Indeed I would go as far as saying the desire for a set of specific normative ethics to impose on others itself equates to the creation of a 'technical elite' in the ethical sphere – an assumed position of knowledge and power. A commitment to the position of uncontrolling and invitational power, a subordinate position in relations of power, is to be sought even in the doing of ethics. This is no easy undertaking. When, as moral beings, we experience the uncertainty and confusion resulting from a lack of ethical norms, then assuming the position of theoretical creator of norms is very attractive: it is much easier to become imaginary experts in a position of knowledge and power than it is to stay with the questions and uncertainty. And it is tempting to assume that I would be right in my choice of ethics, and not to take into account the wide range of views and practices of others. How then can one direction be chosen over another? How can ethical behaviour be developed and ethical norms established in society? From a Christian perspective the answer begins with Jesus, moves from there into communities of faith and then on into involvement by Christians in society.

Jesus as archetype for the subordinate position of power

Jesus approached his engagement with society with a radical commitment to a position of subordinate power. The stories of his temptations relate how he considered and rejected the opportunities to take power that would control others. Many people around hoped and expected that he was going to change tack at some point and reveal himself as the promised Messiah, conquering and liberating them from the occupying forces. They were disillusioned and disappointed as he maintained his position through to his death. When questioned by John the

Baptist's followers as to whether he was or was not the Messiah, Jesus answered by pointing to the invitational aspect of his power – the blind see, the lame walk, lepers are cured, the deaf hear, the dead live, good news is preached to the poor.[7]

In regard to politics, however, Jesus did not assume a passive and accommodating position but was engaged at the tactical level. For example, he teased the country's occupying forces, by advocating walking the 'second mile'. Commentators note that a Roman soldier could ask anyone in an occupied territory to carry their pack for them for one mile – and one mile only (on threat of punishment, because the ruling forces did not want to breed discontent through unreasonable laws). The offer of carrying a second mile relieved the soldier from the task of finding someone else to do this unpopular task, but acceptance carried the risk of being found out and punished. And of course no one would ever believe that someone had *offered* to carry it further! Thus Roman laws would be subverted from within as the soldiers faced the dilemma of what to do and possibly got into trouble over it. Jesus was highly engaged through the same kinds of action I have argued for in the participative approach to shaping technology and cyberspace.

His approach was the same with ethical issues. The Law, he said (referring primarily to the Ten Commandments) would not pass away but was fulfilled by adherence to it in the position of subordinate power. God, the giver of the Law, was not to be challenged, because it was he who gave the Ten Commandments for the purpose of creating a new kind of society with a people who had different relations among themselves. This was to be a sign to the rest of the world. Surprisingly, perhaps, no further sets of rules were required; only a continued radical adherence to those already given. Although the temptation when faced with new situations, confusion and uncertainty, is to jump to the imaginary position of an ethical elite, proposing 'solutions' for the whole of society (that probably cannot reasonably be put into practice), this is to assume the position of lawgiver – which belongs to God alone. While affirming the Law and all it entailed, Jesus nevertheless spoke and acted from a subordinate position, one of obedience to the Law. From that place, he called for re-examination of every area of life: the nature of ownership, the nature of truth, sexual relationships, and so on. To follow Jesus in this way is an enormous challenge that requires honesty and courage. For example, in just one area, what I am doing by holding on to wealth when other people need it would constitute theft.[8] Following such a line, some activities of individuals and organisations, quite legitimate under capitalism, might look rather different. For example, selling software that does not work effectively, easily and reliably, or creating software that requires upgrading of hardware and operating systems, might be considered tantamount to theft.[9]

Jesus pointed to the principles of subordinate power underlying the Law and challenged the Pharisees and Sadducees on their interpretation and misuse as a burden on people instead of liberation into new ways of relating.[10] He challenged them in dialogue: 'You have heard it said, but I say unto you . . .', and by creating controversies, for example healing on the Sabbath and meeting those they deemed unacceptable company. Concentration on their ethical codes of conduct and practice in preference to an understanding and practice of the underlying principles actually blinded them to God, Jesus claimed.[11] We might contemplate whether our keenness to determine ethical codes in relation to new situations might actually blind us to what God is doing, or to how God is revealed to the world through the new technology.[12] Of course, this is not to say that we do not need ethical norms. Specifically they should be developed in a context where others care enough to point out the propensity for self-delusion, where failure is forgiven and where people can go on exploring what it means to be faithful to the Law. This is why church becomes central in exploring morality.

Communities of faith work out morality

I have argued for developing 'networks of concern' in the response to technology; for developing a 'situated knowledge' of a portion of the technology from the perspective of lived experience, and then applying tactical responses. I suggest that the equivalent task in ethics is done within a community of faith (or 'church') and its practices. Hauerwas proposes that the church is the place in which to work out the kind of morality needed in society: 'The church, then, becomes the politics – the dominion – that makes the exhibition of the morality God desires for all people a material reality. To be part of such a politics is to be provided with the means to live the way God created and intended all humans to live.'[13] Within a community of faith there could be formed a body of people committed to the project of establishing a morality appropriate to the situation of twenty-first-century life; of gaining a situated knowledge of that morality; and of engaging in tactical initiatives in society. Such initiatives would not lay claim to any strategic position of knowledge of ethics, but of tactical engagement in provoking moral controversies, in engaging with people in innovative dialogue and initiating participation with the practitioners, and in appropriating existing morality in surprising ways.

There are good reasons why such an endeavour is well placed within the church. First, there are real differences of opinion and thought with which to struggle while knowing that people are still bound by a common goal. Second, there is a commitment to and a demand for truth-telling that extends to finding ways of talking about what happens in society. We consider each in turn.

The process of developing a situated knowledge of ethics is not an easy one and the church is certainly not a cosy place in which to begin. People are gathered from all walks of life and (as I experience in my church) can find themselves on different sides of issues in their life and work. There are rich and poor, black and white, citizens and refugees, those that speak English well and those that stumble, the temporarily enabled and disabled, those with 50 years of faith and commitment, and those passing through. Yet this is exactly why it is the place to engage as a 'dominion in which to exhibit the morality God desires for all people'. Images of the future as presented to us with absence of all conflict look suspect, appearing as illusions conjured up to show a people pacified, not a peace that has been worked out through the humble commitment to one another, bearing differences that do not vanish.

Three images for the future of society as a result of cyberspace are frequently presented and they are the continuation of lineages described in Chapter 2: Bill Gates has presented the image of 'friction-free capitalism', by which he means that the technological revolution is to be adapted into commerce to make it more efficient, and so to facilitate global informational capitalism.[14] The second vision is of a kind of communitarianism – the idealisation of community that will be restored if only everyone communicated, a refuge in a mythic realm of stability and order away from the disorders of change and conflict in the real world.[15] The third projection is the 'global brain', being the transfer of localised and situated knowledge into the disembodied communications space we call cyberspace. But what if the friction is in fact the grip of handshakes and meetings that make business happen? What if the struggle to come to an agreement is part of the building of trust and commitment to overcome problems and not just a cost? 'But what if these [conflicts and antagonisms of the real world] are in fact constitutive of social and political life? What if they are actually the condition of possibility for civic and democratic culture?'[16] But what if the impossible-to-share situated knowledge is exactly that which transforms and leads to creativity and new ideas? What if the differences are what makes the difference? Transforming encounters with people with whom we have conflicting views and the embrace of new ways of seeing and living are of the essence of loving our neighbours.

In the context of a community of faith, there is a general commitment to seeking the truth and to finding ways to avoid self-delusion. Part of this process is finding a language with which to reflect truth accurately. Finding a language with which to talk about ethics in cyberspace is one aspect of this. One result of this is that the approaches that emerge may sometimes look rather different from conventional ones. You will note, for example, that I have avoided the word 'privacy' in so far as I have been able, in order to avoid invoking the whole panoply of the USA Constitution that is attached to it. It is not that I want to circumvent the issues, just that the language is too value-laden to begin a meaningful discussion. Instead I have talked in several chapters about data security, questions of self-disclosure, and protection from abuse by governments etc. In the area of cyberspace there are other examples of language that invoke technical solutions from a position of power that I have also sought to avoid. One such is 'disinter-mediarisation'. Many corporations are introducing customer service support through telephone machines. From their per-spective, they are removing layers that get between them and their customers – a strategy they call 'disintermediarisation' – and it allows them to structure all the conversations, record them on computer, analyse them, so giving them knowledge and the opportunity to make their business more profitable. From the customers' point of view, the lived experience is one of cost: paying for phone calls and not speaking to anyone, time spent waiting, having difficulty in selecting numbers when the particular query does not fit the categories ... To users, it feels as though something has got in the way of communicating with a person in the company – a new 'mediation'. So the first task facing the community seeking to work out morality is the development of ways of talking about what is happening – a key part of truth-telling.

Monsma advocates a programme of combining communities of faith and technical networks of concern, arguing that Christ-ians should be acting to develop situated knowledge about technology themselves.[17] Here I suggest that such work should be based within locally situated, geographical groups – that the church engages with ethical and moral issues from the position of subordinate power and, from there, its members engage in different technical networks of concern, either as individuals or part of other organisations. However, the link between community of faith and network of concern can be strong. Experience by participants of groups exploring ethical issues as part of networks of concern run by the Church of Scotland's Society, Religion and Technology Project, showed that people were drawn (back) into communities of faith as a result. Other networks have found that personal identification with Jesus' subordinate position gives them a language for their experience and so faith, as happened in some of the Greenham Common women's groups. By participating in society in more general networks of concern, one also honours others struggling in the same direction.

Support and distinction from other faiths

Reading the Buddhist Hershock's book on the Information Age, I was struck by how much I agreed with him. Hershock argues that an approach is required that radically alters our perception of technology (which he claims is based on will) and chooses the three marks of Buddhism (*laksaṇa*). These are offered as an alternative to the control and independence which he sees embedded in Information Technology.[18]

Focusing on the *impermanence* of things means recognising both the fragility of our endeavours and also the provisional nature of knowledge – and so consciously looking at them from different perspectives and at different scales. What does it look like in the short term, in the long term? What does it look like from the perspective of a user, from the environment, from employees? He emphasises the energy that comes from pursuing this sense of impermanence – if everything is impermanent, then change is possible. Focusing on *selflessness* means realising no 'thing' (including change itself) is universal and eternal. Nothing stands apart from and separate from anything else. Every action taken affects something else, all is interrelated. Finally, the third mark is that everything can be characterised as *'troubled'* or *'troubling'*.[19] Although we might experience a technology as beneficial, it might be causing 'trouble' somewhere else – the manufacture of computers, for example, is shifted around to find the cheapest source, thus suddenly leaving whole communities out of work, and it uses highly toxic compounds in its work. A corollary of this is that everything is 'out of kilter' and therefore demands our attention. As technology becomes more tightly coupled, with one thing affecting the next and so on, this means that we have to pay more attention to it, not less. Hershock concludes that 'there is no horizon of relevance to us and so there is no limit to our intimacy with and responsibility for things'. He extends his approach to develop a theory for the interrelatedness of all things and the need to express that in an improvisatory and dramatic way.[20]

The Buddhist line here is not directly contradictory to much of Christian thought, nor to the line I am taking. In a pluralist society, religions often find support from each other, as they seek for life beyond consumerism and the experiment of individual humanism. Hershock's approach is presented in the context of a criticism of the church – admitting this was not what its founder meant to happen. He says that it is will-based and controlling. This is exactly the opposite of what Jesus meant – his will was always to do that of his father, not his own; he chose not to control, but stay in a position of subordinate power. Between this Buddhist approach and a Christian one, there is, however, the key difference relating to the person of Jesus. For example, the impermanence of all things is stressed by Jesus (excepting the law and his words); the selfless aspect is stressed by Jesus, but a community of faith is to find identity (a problematical concept in Buddhism) in Jesus and a shared life responding to him; finally nothing stands righteous and without 'sin' before God, so all technology has a sinful dimension, but is to be redeemed by the presence of Christ and his promised second coming. The claim that Jesus makes has to do with the uniqueness and particularity of himself.

For Christianity is not founded on a set of practices, but grounded in the person of Jesus Christ, guided by the Holy Spirit in a community of faith. What does this particularity of Jesus imply in the contexts we have studied, the social spaces changed by cyberspace? What does it mean for information and knowledge, for identity and vocation and the formation of selves, for community and communication, for belonging to a society? In conclusion we look at the connection to Christ as the aspect that transcends and transforms all the others.

Connected to Christ

Jesus and his presence and incarnation present particular challenges in the context of cyberspace. Christians talk of his presence – promised by him when 'two or three are gathered in my name' – but that presence can seem as virtual as the digitised or imagined persona found in cyberspace. In what way is an online mediated presence different from the one Jesus meant? When the Johnson-Lorenzes began an online church in the early 1980s, they argued that for the first time people could worship in spirit and in truth, without being distracted by whether or not the person next to them was beautiful or smelly. Could Jesus fulfil the promise of his presence when two or three are gathered online? The counter-argument focuses on the importance of the body, but it often tends to over-simplification – it runs (to be succinct to the point of caricature): Jesus had a body; therefore bodies are important; therefore we should protect those aspects of the body which appear to be threatened by the introduction of cyberspace, for example face-to-face communication, direct and unconscious knowledge, co-located neighbours etc. What this debate highlights is that the introduction of cyberspace raises some interesting questions on the nature of knowledge of the presence of Jesus and of the place of the body in Christian thought. This is not the place for a full review of these questions, but I will pick up certain aspects that shed light on the issues raised in the book on self, knowledge, communication and community.

First, the body does have a significant place in Christian belief. Jesus is the exemplar of what it is to be truly human – and it is possible to be human with all the limitations, constraints and frustrations of the body; one does not need to escape the body in order to live fully (and obediently to the Law). Jesus also demonstrates that knowing happens through the body, not just through the mind. In this cerebral age where, as we have observed, cyberspace engages the conscious, articulate mind, it is tempting to think that all knowing is acquired in this way. But communication also happens through our body: our postural response in shame, deceit, desire and boldness, for example, are all communicated by body and known by other bodies. Jesus knew when someone had touched him in the middle of a crowd. Knowing can and does happen in ways other than through the conscious mind and so other than by cyberspace.[21] That is one reason why church services are sensual and engage feelings, rather than simply being word-based. The body is also the distance of relationship between God and his people. Jesus lived at the bodily distance of friend, companion, lover, teacher and debater. It is from the distance of a companion or neighbour that Jesus says to his friends that if they have seen him, they have seen God.[22]

Finally, the body of Christ was given a special significance in his appropriation of the Passover meal the evening before his death; he replaced the usual words in the liturgy with 'This is my body given for you, do this in remembrance of me.' His instruction was to remember him by and through his body. To be obedient to this command means physically meeting in order physically to partake of food. The presence that Jesus promised comes as people gather in this way, as affirmed by the New Testament story about the two disciples on the road to Emmaus who recognised him as the food was blessed and they began to eat.[23] The subordinate position of power for the sake of others is recalled face to face with others.

In our Bible Study group, we would find it a difficult

transition to live without the phone and email as these are ways
we keep in touch with each other between weekly meetings, as
well as dropping into each other's homes and meeting in pairs
to talk, pray, comfort and rejoice. The IT is used as an adjunct to
the weekly sitting down together, usually with a meal first,
to recall explicitly what Jesus was like, what he did, and to
listen to words of others and learn how to be called into being
for each other. I believe it is communities of faith like these in
which our identities are formed in relation to Christ, and our
individual stories woven into the history of God. In a personal
narrative, a new identity is formed by choosing to follow Christ
and turning away from past behavioural patterns into new ones
under the Law, formed in the context of a community of faith
engaging with the 'presence of Christ', and called out by others
and for the sake of others.[24] No attempt need be made to create
a 'self' – 'self' forms by living in the world, getting involved
in it from below, responding to others.[25] It is *for* others, unlike
the creation by oneself of the 'expressive self'; it is a life with
a direction, unlike the Protean multiple self, or the consumerist
'plastic self'; and it is a narrative which is endless, always full
of possibility, always moving and changing – one meaning of
eternal life.[26]

Does the form of cyberspace, with which we began the
chapter, help people towards or away from such an identity in
Christ? Or, to frame it as I did there, does the exploration
and serendipitous discovery, discovering what there is to buy,
engaging in play space pretending to be someone else, communicating with new people, leaving bodies behind, the
emphasis on the conscious articulating mind, the silent surveillance and grabbing at our attention – does all this help us
towards identity in Christ and living 'from below'? I cannot
answer for you, only for me; I find the form seductive and it
sometimes leads me away from the engagement with people
around, keen to avoid the bodily tension of conflict; I find the
viewing of options as to what I could buy makes me dissatisfied
when I have plenty; I find it fun to initiate new conversations
rather than to commit time to participate with long-standing

friends, either in the material world or cyberspace. In consequence I am selective about what I use IT for, when I use it, how often. I choose to be focused in seeking information, careful about the time I spend online, admitting a responsibility to myself for my clickstream and, more generally, my data-image; I am selective about what I share, reserving much for friends and church members in face-to-face vulnerable conversation; I seek to avoid being tempted on what there is to buy and so what I lack; to communicate more with those nearby than those far off,[27] attending to them and being response-led as well as demand-led, open to the disagreements and unable to click their bodily presence off.

But, the form of cyberspace is not its final form – for it is malleable and you can participate in shaping it for the well-being of people.

Conclusion and apologia

What I have tried to do in this book is to explain how someone can develop and shape technology, in this case information technology and cyberspace, while following the way of Jesus Christ. Or, to be more specific, how do I express my Christian faith in shaping IT?

First, the task is possible. Cyberspace technologies change the ways people relate to each other. They are moulding social life. They are, however, malleable – shaped by what people choose to use, choose to buy, and choose to attend to. But they don't only come as commodities in the way that jugs or rugs come, with the market simply determining what gets developed. People also appropriate them for their own uses and transform what had been intended. The use of data networks for communication is one example. This illustrates that the task is possible – technology can be shaped.

Second, the task is worthwhile. The detailed investigation of the many places in which there is confusion or doubt about how to behave, what laws to introduce, what can be expected of others' behaviour, illustrates that there is still much to be decided. And I considered only a small part of the impact of cyberspace on our lives. There is an opportunity for getting involved in that decision-making.

How does one bring Christian values into a personal engagement in society? We reviewed several alternatives briefly, ranging from the separation of Christians to create a new society (the Amish) to just acting in society, seeing Christian values as not relating to society at all. These two extremes can be expressed in terms of futility – for example, when people claim that there is no point in attempting to bring Christian values into society until everyone in that society believes in God and so accepts them. This was said to me several times in the course of writing this book. There is deep substance to this view, especially since Jesus' primary task was the re-establishment of a priestly kingdom and holy nation, open to all believers. However, ethical norms associated with IT and cyberspace will be established, irrespective of faith or none, and I would like to see that Christian values are at least presented for consideration. I, for one, am not willing to wait for the time when all recognise God, which is promised in the Bible only at the second coming.

The next question to emerge is to ask what values Christians bring and how they bring them. I have argued that they at least include the way we relate to one another in relations of power (for example in the use of money, sex, attitudes to parent and strangers). The new relationships formed by a radical adherence to the Ten Commandments as taught by Jesus are to be extended by us to engagement with the people we live among, whether neighbour, friend or enemy. This approach always uses a non-controlling and invitational power for the sake of the other. It therefore determines how one goes about introducing the values. It implies the rejection of the mediated approach – persuading or converting the technical elite to your point of view and to the values that you hold ('convert the kings') so that then what matters to you will be enacted through their activities. From that position their activities could be temptingly limiting and controlling. That was not Jesus' way: he chose people like fishermen, and so I have examined how non-technical people, those who experience the results of the development, can influence the shaping of technology. Consequently this book has been specifically addressed to the 'ordinary' person in regard to IT and cyberspace and not to the programmers, managers, investors or politicians who form part of what I call the technical elite and who, in the Western world, take decisions on behalf of others. The book starts from the position of lived experience and seeks to maintain a subordinate position of power to effect change (although I note I have lapsed more than a few times into the mode of 'what I would do if I were in charge' – an approach I criticise).

Christian morality, I have asserted, is formed in communities of faith by people in society engaged in the endeavour of living out new relationships in response to Jesus Christ. The learning I discover there is what I then apply through my life in participating in society. This is not the only way that God is at work in society – the book takes only one strand of possibility. The explicit references to theology, that of neighbour and of incarnation, have been brief, but they are imperative for the church as they consider the formation of their Christian relationships in the context of a twenty-first-century society. Here I have been concerned to illustrate why they are worthy of attention and how they relate to the questions of the time, rather than attending to them directly.

In examining social spaces, I have delved as deeply as I can for an introduction, but am aware that many issues have not been raised – particularly in areas of justice on the worldwide scene. This too can be tackled from 'below' through the pain of non-communication with those who are excluded from the Internet by reason of expense, politics, religion or their government's lack of money because of debt repayments.

Finally, I return to the question I asked in the introduction. Does the teaching of Jesus have anything to say to people beginning the third millennium? My answer is that it is effective. I have tried and been engaged with all the techniques discussed in the participative approach, both in the positions I have chosen in professional work and as a citizen. In my first job as researcher, the group using the electronic service appropriated it in ways unexpected and not totally welcomed by the funding body, although it made a significant step in researching and improving human–computer interaction. I have initiated technical controversy and established innovative dialogues with users. I have participated in government lobbying and forums on shaping technology. Jesus' way works.

In describing them I have had to find a way of talking that does not invoke religious language – this has been incredibly difficult at times. I have been tempted to use words like 'love' many times but resisted. This is my opportunity to say that the non-controlling and invitational approach to people and to shaping cyberspace is one of love for the other in response to Christ's love for us.

Notes

Chapter 1: **Introduction**

1. M. Kaku, *Visions: How science will revolutionize the 21st century and beyond* (Oxford: Oxford University Press, 1998), p. 19.
2. E.g. M. Castells, *The Rise of Networked Society* (Oxford: Blackwell, 1996).
3. E. Nicholson, *Freedom in the Electronic Age: A lecture by Baroness Nicholson of Winterbourne to the John Stuart Mill Institute on 16th November 1998* (London: John Stuart Mill Institute, 1999), p. 16.
4. For example, P. Levy, *Becoming Virtual: Reality in the digital age* (New York: Plenum, 1998); Kaku, *Visions: How science will revolutionize the 21st century and beyond*; N. Negroponte, *Being Digital: The road map for survival on the information highway* (London: Hodder & Stoughton, 1995).
5. For example J. Ellul, *The Technological Society* (New York: Vintage, 1964); P. D. Hershock, *Reinventing the Wheel: A Buddhist response to the information age* (Albany, New York: State University of New York Press, 1999).
6. For example, Ian Barbour's writing; Schuurman and others in the Calvinist tradition; Jacques Ellul, David Lyon and the work of the Society, Religion and Technology Project of the Church of Scotland (see Annotated Bibliography).

Chapter 2: **Locating Information Technology and Cyberspace**

1. For example M. Castells, *The Power of Identity* (Oxford: Blackwell, 1997).
2. K. Robins and F. Webster, *Times of the Technoculture: From the information society to the virtual life* (London: Routledge, 1999).
3. A. L. Friedman and D. S. Cornford, *Computer Systems Development: History, organization and implementation* (Chichester: John Wiley & Sons, 1989, reprinted 1993).
4. Robins and Webster, *Times of the Technoculture*, p. 97.
5. Ibid., p. 99.
6. Ibid., p. 98.
7. H. C. Link, *The New Psychology of Selling and Advertising* (New York: Macmillan, 1932), quoted directly from Robins and Webster.
8. P. D. Hershock, *Reinventing the Wheel: A Buddhist Response to the Information Age* (Albany, New York: State University of New York Press, 1999).
9. M. Castells, *The Rise of Networked Society* (Oxford: Blackwell, 1996).
10. Robins and Webster, *Times of the Technoculture*, p. 91.
11. Data from www.cyveillance.com.
12. J. Carr, anarchy.com. *Prospect* (1999), pp. 22–6.
13. Most European networks used a different protocol. While I was working for the Institute of Physics, we had the first website on JANET, the UK academic network, to be Internet (TCP/IP) only in 1994.
14. P. Levy, *Becoming Virtual: Reality in the digital age* (New York: Plenum, 1998).
15. Ibid.

16. S. Johnson, *Interface Culture: How new technology transforms the way we create and communicate* (London: HarperCollins, 1997), argues that what began as a useful metaphor of how to use computers – the desktop – came to be applied more generally in how we view the world.

17. T. Berners-Lee, *Weaving the Web* (London: Orion Business, 1999).

18. M. Wertheim, *The Pearly Gates of Cyberspace: A history of space from Dante to the Internet* (London: Virago, 1999), p. 227.

19. Ibid., using Tuchman.

20. Ibid.

21. Ibid., p. 229.

22. My MPhil on fuzzy space was initiated by reading about factors in decision-making and thence developing a new mathematics to apply to this kind of dimensioned space.

23. Research on academics wanting help in getting to online journals – a vital resource for their work – said that a 24-hour email response was fine. C. Baldwin and D. J. Pullinger, *SuperJournal: What readers really want from electronic journals* (4th ELVIRA Conference, ASLIB, 1997).

24. Levy, *Becoming Virtual*, p. 31.

25. Ibid.

26. Herbert Burkett, 'The Ethics of Computing?' in J. Berleur, A. Clement et al. (eds), *The Information Society: Evolving landscapes* (New York: Springer-Verlag, 1990), pp. 4–19.

27. The latter identified by Castells, *The Power of Identity*.

Chapter 3: **Technology**

1. M. Castells, *The Rise of Networked Society* (Oxford: Blackwell, 1996), p. 5.

2. N. Negroponte, *Being Digital: The road map for survival on the information highway* (London: Hodder & Stoughton, 1995); M. Kaku, *Visions: How science will revolutionize the 21st century and beyond* (Oxford: Oxford University Press, 1998).

3. F. Stalder, 'The network paradigm: social formations in the age of information', *The Information Society: An International Journal* 14/4, (1998).

4. T. Berners-Lee, *Weaving the Web* (London: Orion Business, 1999).

5. R. Boyce (ed.), *The Communications Revolution at Work: The social, economic and political impacts of technological change* (Montreal and London: McGill-Queens University Press, 1999).

6. S. Turkle, *The Second Self: Computers and the human spirit* (New York: Simon & Schuster, 1984).

7. Melvin Kranzberg, 'The Information Age: Evolution or Revolution?' in B. R. Guile (ed.), *Information Technologies and Social Information* 'Technology and Social Priorities' (Washington, DC: National Academy of Engineering, 1985), pp. 35–54. Some technologies appear to be all good for human beings, but have negative impact on the environment.

But what about those biological weapons? Surely some technologies can be seen to introduce so much 'badness' that despite some good, we would have to consider them evil? I think that weapons of mass and indiscriminate destruction are a special case, but what if there were information technologies like that, for example for total surveillance of all speech and movement of every single member of society, as in Orwell's *1984*? In Britain we hold back from connecting the government databases one to another to protect against this idea, and resist the introduction of an identity card. But in the year this is published, GPS chips can be implanted in everyone so that no one gets lost, we know where children are, and we can be guided in strange places – and where we are can be known at all times. New technologies are introduced because each new step contains 'good', and we are tempted by the 'good' into adopting them, even though we know some effects are limiting or harmful.

8. I am following my entry: D. J. Pullinger, article 'Technology' in José Míguez Bonino (ed.), *The Ecumenical Dictionary* (Geneva: World Council of Churches, 1989).

9. This was widely reported in the newspapers, but I have been unable to locate a reference for the reader.

10. For example ITL, who designed and built the Diamond computer which was in widespread use by media firms, government and commercial organisations in the years around 1980.

11. D. Milsted (ed.), *They Got it Wrong! The Guinness dictionary of regrettable quotations* (Enfield, Middlesex: Guinness, 1988), pp. 110 and 202.

12. E. Tenner, *Why Things Bite Back: Technology and the revenge effect* (London: Fourth Estate, 1996).

13. J. Gleick, *Faster: The acceleration of just about everything* (London: Random House, 1999), pp. 217–25.

14. Several alternative interpretations of his outrage could be made but, in the context of CFCs, I feel he was not arguing that there is an interventionist God but that humans are allowed to proceed with too little knowledge relative to the damage that we cause and that is, at some level, 'unfair'.

15. S. Hauerwas, *Sanctify Them in the Truth* (Edinburgh: T. & T., Clark, 1998).

16. P. Ricoeur, *Oneself as Another* (Chicago and London: University of Chicago Press, 1992).

17. M. Castells, *The Power of Identity* (Oxford: Blackwell, 1997), pp. 333–42.

18. I. Barbour, *Ethics in an Age of Technology: The Gifford Lectures* (London: SCM Press, 1992).

19. Ibid.

20. Unattributed research quoted by John Cassidy in the *New Yorker*, 27 November 2000.

21. R. Dahrendorf, *Class and Class Conflict in Industrial Society* (Stanford, California: Stanford University Press, 1959).

22. It is in fact nearly impossible to test all the features and the interface to them. Indeed some software states are only met after years of use, and so if there are bugs, then the purchaser and/or user will not find them until much later.

23. R. Bogoslaw, *The New Utopians: A study of system design and social change* (Englewood Cliffs, New Jersey: Prentice-Hall, 1965).

24. Exodus 19:6.

25. Compare 'The Truth about God' in Hauerwas, *Sanctify Them in the Truth*.

26. David Sanders, my New Testament teacher, introduced me to this interpretation of the commandments.

27. S. V. Monsma (ed.), *Responsible Technology: A Christian perspective* (Grand Rapids, Michigan: William B. Eerdmans, 1986), pp. 68–9.

28. Here I am following D. F. Ford, *The Self and Salvation* (Cambridge: Cambridge University Press, 1999) in his engagement with the ideas of Ricoeur in the ethical demand of the Other. I am being deliberately vague about who constitutes a neighbour here; I address the topic in Chapter 5 and, at the very least, one could take a Barthian line and be concerned for the neighbour who is a member of the community of faith. Karl Barth, *Church Dogmatics*, Vol IV, Part 3, 68, *The Holy Spirit and Christian Love*.

29. E. Davis, *TechnGnosis: Myth, magic and mysticism in the age of information* (London: Serpent's Tail, 1998).

30. For a summary of the five positions see C. Mitcham and J. Grote, *Theology and Technology: Essays in Christian analysis and exegesis* (Lanham, MD: University Press of America, 1984). Mitcham claims that Niebuhr himself seems drawn to the conversionist position. In taking this position, I do not want to suggest that it is the only or primary way that Christ is present in the on-going activities of the world, but is a helpful way of stirring myself up. In particular I am not adopting a Deistic position – one in which God created the world and then left us to get on with it.

31. This position is similar to that of the Calvin Center for Christian Scholarship (Monsma (ed.), *Responsible Technology*), in which they distinguish a collective responsibility that every person shares for the technology we have (p. 225), and the role-specific responsibility of, say, a computer programmer, or marketing manager.

Chapter 4: **Information and Knowledge**

1. There are some interesting questions as to the limit of what should be sold, but I do not discuss them here.

2. A. Borgmann, *Holding onto Reality: The nature of information at the turn of the millennium* (Chicago: University of Chicago Press, 1999). The role of intelligence is important to Borgmann who identifies five terms that are required in order to produce information: 'INTELLIGENCE provided, a PERSON is informed by a SIGN about some THING within a certain CONTEXT.'

3. This was reported in the *Yorkshire Evening Press* (14 October 1998), and I read it in the archives of Association for Computer Machinery's *The Risks Digest*, available at URL catless.ncl.ac.uk/Risks.

4. Reported in the *New Scientist*.

5. Introduced by a corporation frustrated by the wage cost in understanding old code.

6. A. Ayckbourn, 'All the world's a stage poorer for the mobile phone', *The Guardian*.

7. Plato, *Phaedrus* 275A–B, quoted in Borgmann, *Holding onto Reality*, p. 48.

8. J. Everard, *Virtual States: The Internet and the boundaries of the nation-state* (London: Routledge, 2000), p. 122.

9. Borgmann, *Holding onto Reality*, p. 15.

10. These were some of the issues raised when I represented a company seeking to help knowledge management in a large corporation which had just merged in 1999.

11. The Knowledge Society is a term used to describe a society in which creating, sharing and using knowledge are key factors in prosperity and well-being of people. The concept of knowledge workers was originally introduced by Peter Drucker in 1959 and his article in *The Atlantic Monthly*, 'The Age of Social Transformation', covers it in more depth. See also Department of Trade and Industry, *Our Competitive Future: Building the Knowledge-Driven Economy*, London (1998).

12. iqport.com 1999–2000.

13. OK, so you heard a different version of it! Some of the versions involve Henry Ford, others remain anonymous. I thought I would stick to the latter, having no way to verify the former.

14. J. C. R. Licklider, 'A crux in scientific and technical communication', *American Psychologist* 21 (1966), 1044–51.

15. T. Berners-Lee, *Weaving the Web* (London: Orion Business, 1999), p. 202.

16. R. N. Stichler and R. Hauptman (eds), *Ethics, Information and Technology: Readings* (Jefferson, North Carolina: McFarland, 1998).
17. D. Shenk, *Data Smog: Surviving the information glut* (London: HarperCollins, 1999).
18. Gallup and the Institute of the Future; I believe the data refers to managers in corporations.
19. T. H. Davenport, and J. C. Beck, 'Getting the attention you need', *Harvard Business Review* 78/5 (2000), 118–26, at 120. Factors not related to attention impact of messages are: if the information comes from a superior; if the content was new or unusual; and whether or not the recipient agreed with the sender about the content.
20. Cyveillance data, from www.cyveillance.com.
21. Davenport and Beck, 'Getting the attention you need'.
22. The professional society to which I belong – the Ergonomics Society – began in World War 2 after it was found that soldiers who spent long days sitting in tanks in the middle of the desert monitoring fuzzy radar screens would increasingly 'see' blips that were not there. It was proved later that this is a normal psycho-physiological response – after the execution of some of those soldiers for traitorous ill-intent.
23. Davenport and Beck, 'Getting the attention you need'.
24. Ibid.
25. Ibid., p. 125.
26. G. Claxton, *Hare Brain, Tortoise Mind* (London: Fourth Estate), p. 130.
27. Ibid.
28. Ibid.
29. Davenport and Beck, 'Getting the attention you need', p. 125. See also www.attentionscape.net.
30. Claxton, *Hare Brain, Tortoise Mind*, p. 131.

Chapter 5: **Patterns of Relating**

1. A. A. L. Reid, 'Comparing Telephone with Face-to-face Contact' in I. de Sola Pool, *The Social Impact of the Telephone* (Cambridge, Massachusetts and London: MIT Press, 1977), pp. 386–412.
2. P. Ricoeur, *Oneself as Another* (Chicago and London: University of Chicago Press, 1992).
3. S. R. Hiltz, *Online Communities: A case study of the office of the future* (Norwood, New Jersey: Ablex Publishing, 1984).
4. For example the research described in Rebecca G. Adams, 'Friendship Patterns among Older Women' in Jean M. Coyle (ed.) et al., *Handbook on Women and Aging* (Westport, Connecticut: Greenwood Press, 1997), pp. 400–17.
5. Karen Walker, ' "Always there for me": Friendship patterns and expectations among middle- and working-class men and women'. *Sociological Forum* 10/2 (1995), 273–96;. Janos V. Botschner, 'Reconsidering Male Relationships: A social-developmental perspective' in Charles W. Tolman, Frances Cherry et al. (eds), *Problems of Theoretical Psychology* (North York, Ontario: Captus Press, 1996), pp. 242–53.
6. N. Postman, *Technopoly: The surrender of culture to society* (London: Vintage Books, 1993).
7. R. Williams, *Keywords: A vocabulary of culture and society* (London: Fontana Press, 1976; revised edn Flamingo, 1983).
8. G. Graham, *The Internet: A philosophical enquiry* (London: Routledge, 1999), Chapter 7, 'New Communities'.

9. Ibid., p. 142.
10. N. Watson, 'Why We Argue about Virtual Community: A case study of the Phish.Net fan community' in S. G. Jones (ed.), *Virtual Culture: Identity and communication in cybersociety* (London: Thousand Oaks; New Delhi: Sage Publications, 1997), pp. 102–32.
11. Ibid., p. 119.
12. Graham, *The Internet*, p. 142.
13. Using the phone means that as someone has not perceived the other's opinion as clearly, they will be more likely to change their own opinion. Perhaps the effect of having time to think, or the distance that gives perspective, suppresses the automatic reactions we have towards people.
14. Reported at a conference in Sweden in 1984, by Starr Roxanne Hiltz. She discovered that, compared to online, people meeting face to face defer to senior males – even in the USA (to her complete surprise).
15. Roel Vertegaal, personal communication, 27 September 2000.
16. Watson, 'Why We Argue about Virtual Community', p. 108.
17. Murray Turoff and Starr Roxanne Hiltz in the early 1980s.
18. We did this in the BLEND project in order to allow people with different speed computers to keep up with online conferences both synchronously and asynchronously.
19. Watson, 'Why We Argue about Virtual Community', has this detailed study of the Phish.net community.
20. I am following J. Locke, *The De-Voicing of Society: Why we don't talk to each other anymore* (New York: Simon and Schuster, 1999), p. 27f.
21. S. G. Jones, 'The Internet and its Social Landscape' in *Virtual Culture: Identity and communication in cybersociety*, pp. 7–35.
22. This becomes increasingly important if the future of companies follows a networked 'shamrock' model of business, as described by Charles Handy in *The Age of Unreason* (London: Century Business, 1989; 2nd edn 1991), pp. 72–92.
23. Proverbs 27:10.
24. Luke 10:29–37.
25. A christological interpretation would imply that Jesus was the neighbour who helped in time of need and therefore is to be loved even if he does not fit in with Pharisaical expectations.
26. Other forces in society, e.g. fear of involvement or of litigation, also means that people are more likely to use IT to call for mediated help, than to do so directly.
27. I have suggested elsewhere that this is a working out of 'bad things happen to bad people'. See J. Reid, L. Newbiggin, and D. Pullinger, *Modernity, Postmodernity and Christian*, Lausanne Committee for World Evangelization Occasional Paper No 27 (Edinburgh: Handsel Press, 1997).

Chapter 6: **Identity and Self**

1. O. Sacks, *A Leg to Stand On* (New York, Touchstone Books; revised edn 1991).
2. P. Virilio, 'Speed and information: cyberspace alarm', *CTheory: International Journal of Theory, Technology and Culture* (1995). I believe this is to be the earliest reference to this idea in his writing.
3. C. G. Jung, *Two Essays in Analytical Psychology* (London: Routledge, 2nd edn 1992), par. 305.
4. M. Volf, *Work in the Spirit: Toward a theology of work* (Oxford and New York: Oxford University Press, 1991).

5. Volf's suggestion is that we could focus instead on the characteristics of identity created with and for others – much as I did at the start of the chapter. Through these, we could seek our meaning and purpose, wherever we are or whichever task we are currently engaged with, as an expression of living.

6. P. D. Hershock, *Reinventing the Wheel: A Buddhist response to the information age* (Albany, New York: State University of New York Press, 1999).

7. D. Lyon, *Jesus in Disneyland: Religion in postmodern times* (Cambridge: Polity Press, 2000), p. 42.

8. B. J. Pine II, and J. H. Gilmore, *The Experience Economy: Work is theatre and every business a stage* (Boston, Massachusetts: Harvard Business School Press, 1999).

9. Lyon, *Jesus in Disneyland*.

10. R. J. Lifton, *The Protean Self: Human resilience in an age of fragmentation* (Chicago: University of Chicago Press, 1995).

11. S. Turkle, *Life on the Screen: Identity in the age of the Internet* (London: Orion, 1996).

12. S. Turkle, 'Looking toward cyberspace: beyond grounded sociology', *Contemporary Sociology* 28/6, 643–8, at 645.

13. Turkle, *Life on the Screen*, p. 322.

14. R. Williams, *Lost Icons: Reflections on cultural bereavement* (Edinburgh: T. & T. Clark, 2000), Chapter 1: 'Childhood and Choice'.

15. Turkle, 'Looking toward cyberspace' p. 645.

16. Turkle, *Life on the Screen*, p. 262.

17. S. G. Jones, 'The Internet and its Social Landscape' in *Virtual Culture: Identity and communication in cybersociety* (London: Thousand Oaks; New Delhi, Sage Publications, 1997), pp. 7–35.

18. Turkle's examples are nearly all of this type.

19. I refer here to the research discussed in Chapter 5. Note that the use of email for distribution of information is not included here as part of communication.

20. A US journalist's online love affair, widely read, in which she fell in love and then fell out in less than a year and wrote the whole story.

21. P. Bromberg, 'Speak that I may see you: some reflections on dissociation, reality, and psychoanalytic listening', *Psychoanalytic Dialogues* 4/4 (1994), 517–47, quoted in Turkle, 'Looking toward cyberspace'.

22. Turkle, *Life on the Screen*.

23. Turkle, 'Looking toward cyberspace', p. 646.

24. Lyon, *Jesus in Disneyland*, p. 95.

25. M. Castells, *The Power of Identity* (Oxford: Blackwell, 1997), pp. 8–9.

26. Walt Whitman uses this as part of dividing his being into three, the self, the me-myself (meaning in some sense a 'real me'), and the soul. I use the term to imply the 'real me' without wishing to use Whitman's structure.

27. D. Lyon, *The Electronic Eye: The Rise of Surveillance Society: Computers and social control in context* (Cambridge, UK: Polity Press, 1994).

28. I designed this feature into the *Nature* website; I believe myself to be the first to do so for a research journal and magazine, and I carefully coupled it with a strong privacy statement.

29. Pressure by the Commissioner for Data Privacy.

30. Even answering perfectly politely and truthfully, that I had not purchased a licence because I deemed that I did not have to do so, seemed to irritate them into further demands!

31. C. Middleton, 'Ethics man: Chris Middleton talks to Simon Rogerson, the UK's first professor in computer ethics', *Business and Technology* (1999), 22–7.
32. Herman T. Tavani, 'Bibliographic listing of work on computer ethics'. URL www.rivier.edu.
33. J. Everard, *Virtual States: The Internet and the boundaries of the nation-state* (London: Routledge, 2000).
34. K. J. Gergen, *The Saturated Self: Dilemmas of identity in contemporary life* (New York: Basic Books, 1991).
35. P. Ricoeur, *Oneself as Another* (Chicago and London: University of Chicago Press, 1992), p. 329.
36. D. J. Hawkin, *Christ and Modernity: Christian self-understanding in a techno-logical age* (Waterloo, Ontario: Wilfrid Laurier University Press, 1986), p. 188.
37. See also D. F. Ford, *The Self and Salvation* (Cambridge: Cambridge University Press, 1999), pp. 90–4 for a broader discussion.

Chapter 7: **Political Engagement**

1. N. Negroponte, *Being Digital: The road map for survival on the information highway* (London: Hodder & Stoughton, 1995), p. 236.
2. J. Everard, *Virtual States: The Internet and the boundaries of the nation-state* (London: Routledge, 2000), p. 7.
3. G. Delanty, 'Self, other and world: discourses of nationalism and cosmopoli-tanism', *Cultural Values* 3/3 (1999), 365–74, argues that nationalism has only thrived because of the lack of alternatives that have been put forward and so nationalism has monopolised the pathos of solidarity, commitment and community. His answer is in a new civic cosmopolitanism. Theologians like Barth were against cosmopolitanism and internationalism on the basis that people should first attend to the particularity of where they find themselves.
4. Castells' trilogy, in particular, the second book: M. Castells, *The Power of Identity* (Oxford: Blackwell, 1997).
5. Castells, *The Power of Identity*, p. 304.
6. Ibid.
7. C. Offe, *Modernity and the State: East, West* (Cambridge: Polity Press, 1996), quoted by G. Graham, *The Internet: A philosophical enquiry* (London: Rout-ledge, 1999), p. 85.
8. This is the ethical line of freedom that Barbour elaborates and that we will pick up in Chapter 8. I. Barbour, *Ethics in an Age of Technology: The Gifford Lectures* (London: SCM Press, 1992).
9. L. Lessig, *Code and Other Laws of Cyberspace* (New York: Basic Books, 1999); K. Robins, and F. Webster, *Times of the Technoculture: From the information society to the virtual life* (London: Routledge, 1999); P. D. Hershock, *Reinventing the Wheel: A Buddhist response to the information age* (Albany, New York: State University of New York Press, 1999).
10. M. Stefik, *Internet Dreams: Archetypes, myths and metaphors* (Cambridge, Massachusetts: MIT Press, 1996); *The Internet Edge: Social, technological and legal challenges for a networked world* (Cambridge, Massachusetts: MIT Press, 1999).
11. Lessig, *Code and Other Laws of Cyberspace*, Chapter 1.
12. L. Winner, 'Who Will We Be in Cyberspace?' in P. A. Mayer (ed.), *Computer Media and Communication: A reader* (Oxford and New York: Oxford University Press, 1996), p. 215.

13. M. Poster,.'Cyberdemocracy: Internet and the public sphere', online article (1995). www. hnet. nci. edu/mposter/writings/democ. html.

14. Graham, *The Internet: A philosophical enquiry*, p. 81.

15. Lessig, *Code and Other Laws of Cyberspace*.

16. B. S. Noveck, 'Transparent space: law, technology and deliberate democracy in the information society', *Cultural Values* 3/4 (1999), 472–91.

17. J. S. Fishkin, *The Voice of the People: Public opinion and democracy* (New York: Yale University Press, 1997).

18. R. Hurwitz, 'Who needs politics? Who needs people? The ironies of democracy in cyberspace', *Contemporary Sociology* 28/6 (1999), 655–61, at 658.

19. Castells, *The Power of Identity*, p. 319.

20. Hurwitz, 'Who needs politics? Who needs people?'

21. Castells, *The Power of Identity*.

22. Ibid., p. 350.

23. A. Feenberg, *Questioning Technology* (London: Routledge, 1999), p. 131.

24. Winner, 'Who Will We Be in Cyberspace?', p. 217.

25. Ibid., p. 216.

26. Feenberg, *Questioning Technology*, p. 101.

27. J. Ellul, *The Technological Society* (New York: Vintage, 1964), pp. 274ff.

28. We should note that the main alternative to this approach is localisation of technology and its technical infrastructure. Localised production and consumption together with services mean that, for example, one might have local computer networks which are connected with everyone else through Internet, much as Hull maintained with their telephone network, while the rest of the UK was nationalised and then privatised. Illich, Castells' construction of local states, and Sclove are all advocates of this approach, but it requires the action of precisely those who are part of the historical technocracy and so seems to be less an immediate path of action than the user-participative approach I am arguing for.

29. Feenberg, *Questioning Technology*, p. 139.

30. Ibid., p. 120.

31. Dr Ian Wilmut was a member of a group in the Church of Scotland's Society, Religion and Technology Project, whose report was published as D. Bruce, and A. Bruce, *Engineering Genesis* (London: Earthscan, 1999). As it happens, this was the first paper from the journal *Nature* which was published on Internet in advance of the print with a commentary to give it context (and I published it).

32. J. Weizenbaum, *Computer Power and Human Reason: From judgment to calculation* (Oxford: Freeman, 1976).

33. I use this approach whenever I can, particularly with large international services, using a small group approach that is, for example, formalised in Dynamic Systems Development Method (DSDM). This is not, however, a foolproof method for product development to make money – the classic case is of products for the aged or disabled which are usually developed in this way, but the final product does not sell, because it looks like a product aimed at the market that has a stigma attached to it. There needs to be a broader engagement in product design, as we noted in Chapter 3; 'function' is not the only criterion in technology.

34. 'Appropriation' derives largely from the writings of the Russian Bakhtin, writes J. V. Wertsch, *Mind as Action* (Oxford and New York: Oxford University Press, 1998). It is a translation of the word *prisvoenie*, the process of making something one's own; there is a sense of Other about what is taken

in, so that there is always a new discovery and a resistance that is worked through (pp. 53–4).

35. I am following Feenberg in this; he cites Epstein on pp. 127–8, 141.

36. Feenberg, *Questioning Technology*, p. 146.

37. Winner, 'Who Will We Be in Cyberspace?', p. 217. He concludes with a call for computer professionals to take the lead in initiating public debate.

Chapter 8: **Extra-connected Living**

1. The totalitarian charge against Christianity is refuted by J. R. Middleton and B. J. Walsh, *Truth Is Stranger than It Used To Be: Biblical faith in a postmodern age* (London: SPCK, 1995).

2. S. V. Monsma (ed.), *Responsible Technology: A Christian perspective* (Grand Rapids, Michigan: William B. Eerdmans, 1986), p. 225.

3. K. Sale, *Rebels Against the Future* (New York: Addison Wesley, 1995).

4. Ibid., p. 158.

5. A. Feenberg, *Questioning Technology* (London: Routledge, 1999), p. 197.

6. L. Gringas, 'Psychological self-image of the systems analyst', *Proceedings of the 14th Annual Computer and Personnel Research Conference, Alexandria, USA* (29–30 July 1976), 121–32.

7. Luke 7:22.

8. S. M. Hauerwas, and W. H. Willemon, *The Truth About God: The Ten Commandments in Christian life* (Nashville, Tennessee: Abingdon Press, 1999).

9. Microsoft, as one example of a leading manufacturer, was quoted as saying: 'If we hadn't brought your processor to its knees, why else would you get a new one?' W. W. Gibbs, 'Taking computers to task', *Scientific American* 2777 (1997), 64–71. With thanks to Harold Thimbleby for this example.

10. Luke 11:46.

11. Matthew 23:13–36: the seven woes pronounced on the teachers of the law and the Pharisees.

12. With thanks to David Leal for formulating the idea in this form.

13. S. Hauerwas,'The Truth about God' in *Sanctify Them in the Truth* (Edinburgh: T. & T. Clark, 1998), p. 57.

14. B. Gates, *The Road Ahead* (New York: Viking Penguin, 1995).

15. K. Robins, and F. Webster, *Times of the Technoculture: From the information society to the virtual life* (London: Routledge, 1999), p. 228.

16. Ibid., p. 231. A good example of the Church engaging in ethical controversy was when prayers for the Argentinian dead were offered in the Service of Thanksgiving over the Falklands War in 1982.

17. Monsma (ed.), *Responsible Technology*.

18. P. D. Hershock, *Reinventing the Wheel: A Buddhist response to the information age* (Albany, New York: State University of New York Press, 1999), p. 106. He says he is not making ontological or existential assertions here, but 'following teachings instructs us to see things as impermanent, self-less and troubling.'

19. The word *dukkha* is derived from a wheel with its axle-hole off-centre, and had, Hershock claims, been unhelpfully translated as 'suffering'.

20. This is a theme that is developed in D. F. Ford, *The Self and Salvation* (Cambridge: Cambridge University Press, 1999).

21. Metropolitan Anthony Bloom and Dallas Willard both claim that God can be heard more effectively in this un-conscious listening, e.g. Anthony Bloom, *School for Prayer* (London: Darton, Longman & Todd, 1970), pp. 60–1; Dallas Willard, *Hearing God* (London: Fount, 1999), p. 193.

22. John 14:9. This is what God is like, not a Protean form of shape-changing,

but one form that is perceivable – Jesus in his body. This is further signified by his permanent incarnation in the glorified humanity of the risen Christ as the vehicle of knowledge of God.

23. Luke 16:30. I am not saying this is the only time that Jesus' presence is manifest, but that this is certainly one point.

24. For example, Colossians 3 goes through this.

25. Bonhoeffer, quoted in Ford, *The Self and Salvation*. p. 253f.

26. This form and shape of a life is described by Rowan Williams as 'soul'. (Talk at Greenbelt Festival 2000.)

27. Cf. Barth in his near and far neighbouring: Karl Barth, *Church Dogmatics* Vol III, Part 4, 54 section 3 'Near and distant neighbours'.

Glossary

Asynchronous: literally, not at the same time. Use refers to when a person receiving a communication does so later than it is sent: telephone conversations are synchronous, answer-phones and email asynchronous.

Browser: a computer program that allows you to access the Web: running a browser enables you to download and view web pages. Examples are Internet Explorer, Netscape and Opera.

Chatroom: places, on the Internet, to communicate with other people who are online at the same time, by sending and reading typed messages. The system used is Internet Relay Chat (IRC).

Clicking / clickstream: in a graphical interface, refers to the act of selecting a visual icon or link and using the mouse or keyboard to 'click' on it to effect some action. In WWW, this activates the link and a clickstream is the sequence of links followed.

Commodification: the process of turning objects or services that are not usually sold and bought into marketed and traded items – thus creating commodities. For example software was originally exchanged free of charge, but corporations packaged and sold it.

Computer conferencing: people communicating via an online 'conference room'. People can enter and see if anyone else is online – and communicate either at the same time or by leaving messages to be read later. A more sophisticated version of a chatroom with functionality to enable someone to chair a discussion, for example. This was one of the earliest forms of online communication, developed in the 1970s.

Cyberspace: the space that users mentally construct when going online and accessing web pages, email etc. Also used to mean the sum total of all computers and all their links (land, microwave, satellites and radio) with all their stored digital data.

Data: measures or facts; also the symbols used by computer programs.

Data network: a set of computers linked and able to talk to each other in order to transfer data.

Data Protection Act 1998: the UK legislation, conforming to European Community legislation, that seeks to limit the use and transfer of personal data. A person whose data is under the legislation is called a 'data subject'.

Databank: a database where you can deposit and retrieve data, analogous to a money bank. An example is GenBank, a public database where scientists deposit gene sequences they have found and others can access them.

Database: a structured collection of data which can be retrieved and accessed or processed by software programs running on it.

Data-image: the accumulated data in cyberspace that is related to a particular person. The data is usually scattered over many computers. The DPA protects against wholesale integration of data.

Digital signature: an encryption of a message that uniquely identifies it as coming from one person – and designed so that any attempt to change the message can be easily detected.

Digital TV/radio: the broadcast or online (including cable) transmission of TV and radio in digital form, instead of analogue.

Direct-knowing: knowing by acquaintance rather than by theory about something 'elsewhere and elsewhen'. This is Russell's distinction between direct and indirect knowing.

Dotcom companies: the colloquial name for companies which only have an existence online, rather than in 'clicks and mortar'. Named after their URL addresses that usually end in .com.

E-business/e-commerce/cyberbusiness: business done using the Internet – this can be anything from an individual buying a book online using a credit card to big corporations dealing with suppliers.

Email: the computer version of the postal service. You can send an email to anyone if you know their email address. Like post, it is 'asynchronous' – the receiver does not need to be there at the same time in order to receive your email.

Encryption: a process of using cryptography to ensure that only certain people can access data and messages. Public key cryptography enables two people who have not met to exchange messages that only they can unlock using a 'key'.

Global informational capitalism: the term used by Castells to describe how corporations are extending globally in a capitalist economy and are enabled to do so by information technology.

Interface/graphical interface: anything that sits between the user and a computer and allows communication and interaction to take place. This includes the screen, a pointing device (mouse), icons, commands, sounds, body movement, etc.

Internet: the global collection of computer networks linked using the Internet Protocol.

Internet Protocol: the technical protocol that governs how computers send packets over the network. The transmission control protocol (TCP) describes how a computer can send a stream of information by breaking it into packets and re-assembling at the far end. Together they are referred to as (TCP/IP).

Internet Service Provider (ISP): literally, the company that provides you with a service that connects you to the Internet.

Knowledge economy: often, also, 'global knowledge economy'. Used to describe the move from an industrial economy to one based on the value of what people know and the ideas they have that generate capital for their employers.

Knowledge society: the society that is based on a knowledge economy.

Mediated communication: communication that passes through machines, organisations or other people before reaching the person for whom it was intended.

Meta-narrative: in this book, as used by sociologists to describe a higher or broader description of what is happening in the world than events.

MUD: multi-user domain (developed from Multi-User Dungeons and dragons). Online games areas where people go and usually invent their own character in order to play.

Netiquette: accepted conventions for email and online chat, including how to address people, end messages and give hints about the emotion you would like to convey in a sentence ('emoticons').

Networks of concern: groups of people forming around a particular concern in order to respond to it.

Online: usually applied to being connected to any distant computer using telecommunications and, in particular, to the Internet.

Postmodern: a description of society in an era after the 'modern'; characterised by a number of viewpoints and particularly by the denial of any single overarching authority or interpretation.

Public domain: applied to material that anyone is legally allowed to copy. This might be something old, which has passed out of copyright, or something which is not considered to be copyrightable.

Realtime: the time of the material world; often used in contradistinction to synchronous.

Scientific management: the principles introduced by F. W. Taylor to streamline production. This was achieved by separating planning and thinking from those actually doing the work and then monitoring and managing them in it.

Search engine: services on Internet that search websites and other information stores and enable the user to put in words to locate items of interest to them.

Server: a program that manages shared access to a centralised resource, such as a website or email.

Situated knowledge: direct knowledge that results from a particular situation as experienced by the person.

Social space: used in this book to refer to the space where we have social activities with others – communicating, forming friendships and community, and participating in democratic processes.

Software: the programs on a computer that allow a user to accomplish tasks, for example, word-processing software, spreadsheet software, email software, a web browser, etc.

Source code: the code, written by programmers, that makes programs perform their functions. Most programmers now write application code, which is like building out of complex Lego components; the source code is like the design of the bricks, circuits and motors themselves.

Strategic: careful planning in overall or long-term operations towards an advantage; used in this book to describe the technical experts' development of technology.

Tactical: finding appropriate action, usually short-term, for the circumstances; used in this book to describe non-technical response to technological development.

Telecommunication: any communication at a distance (coming from the Greek *tele* meaning 'far' or 'distant').

Tele-presence: the 'presence' of a person who is physically at a distance.

Tightly-coupled (system): systems in which slack is removed to maximise use of equipment and/or people but often with the result that a problem in one area will have knock-on effects in many others.

URL: Uniform (or Universal) Resource Locator. The addresses for Web pages that allow people to find and link to them.

Video conferencing: a group of people communicating at a distance by means of moving picture and sound.

Virtual reality: the replication of reality or creation of an imaginary reality in cyberspace. In particular, when the body interacts with all-round vision, sound, and sometimes exoskeletal devices – all of which can act upon you and be acted upon by you in cyberspace.

Virtualisation: the process of moving things into cyberspace; applied, for example, to creating libraries of digital content or constructing a lifelike environment in which to train pilots to fly new aircraft.

WAP: Wireless Application Protocol. The technical specification that Internet service uses to communicate with wireless devices, such as digital mobile phones.

Web browser: *see* Browser.

Web pages: Web pages are documents viewable by a Web browser. They are analogous to the pages of books, magazines and directories and can include text, pictures, sound, videos and links to other web pages. Each web page has a separate URL.

Webmail: the use of email through a web browser.

Website: a collection of web pages, usually with one aim or owner, like a book or magazine. They often have entry points to facilitate navigation to the web page required.

Western world: North America and Western Europe; in particular, their culture and social organisation.

Wireless: any device that uses WAP. Also in USA, a digital mobile phone (and in UK the old usage of wireless meant a radio).

World Wide Web: the application on Internet, invented by Tim Berners-Lee, that enables people to build web pages with links to others and then for users to view and follow the links.

Zero-sum-activity: any activity where, in doing it, you cannot do another. Often applied in the context of time and, in this book, of attention.

Select Annotated Bibliography

Technology

There are many approaches to thinking about technology; to do so effectively one must at least be able to explain why current debates arise. I found Feenberg (1999) particularly helpful in this regard. Anthologies of writing can be a good place to start and Teich (1997) contains excerpts from many of the writers cited in this book. As an introduction to how technology is introduced and developed, Tenner (1996) gives detailed explanations of why technology has unforeseen consequences and why they are inevitable (but of course not what they are!). Accounts of the introduction of earlier technologies can help to put the issues into context; the beginnings of the telephone, for example, are chronicled by de Sola Pool (1977).

There are few books explicitly on technology from a Christian perspective, the majority of writing being either in articles or on the related subject of science. The Calvin Center writers have been most prolific and worked together in Monsma (1986) to produce a key book: its main thesis is the idea of differing responsibilities people have as a result of their roles in society ('position-specific responsibility') and as members of society ('collective responsibility'). Ian Barbour has been a significant individual worker over the decades, the latest contribution being Barbour (1992). And Hawkin (1986) has explored the relationship between Christianity and technology, concluding that the Christian view is that technology is permitted, rather than to be actively pursued.

Information society

The major sociological attempt at a coherent explanation and analysis of the underlying thrusts of information society is by Castells in his trilogy; I particularly recommend the second (Castells 1997). From a politico-economic perspective, Robins and Webster (1999) argue that the main purpose of IT is to maintain the status quo for the dominant players – the large commercial and military organisations. This is considered from another angle by Mosco (1989), in describing the paradigm shift of commodification of much of life, and developed further in an edited collection (Mosco and Wasko 1988). Hershock (1999), in a similar analysis sees the underlying dynamic as lying in our will-based approach to life, with a possible transformation through interplay between people who are present to each other. Graham (1999) presents a short and very readable book, which demonstrates both the value of philosophical enquiry for Internet and presents ideas and lines of argument to pursue.

More focused approaches are found in other books, such as Levy (1998), where he argues for the continuity of technological development in the direction away from the specificity of the 'here and now' into shared and usable spaces of 'virtualisation'.

Some of the most useful reading comes from collections of articles or selections, which, if they do nothing else, indicate the very wide range of areas of life which this technology touches, and the different approaches that are possible to consider its impact. The best reader for the background to this book is Mayer (1999). It is divided into two sections: history and systematic studies. The former contains key articles ranging from Vannevar Bush's 1945 article introducing the idea of a personal machine for navigating information, through Alan Turing on computing (1950), and Licklider, one of the prescient writers of the 1960s on information, to Ted Nelson in new systems (1982). The systematic studies section contains key articles from Steven Jones on virtual communities (1995); Allucquere Rosanne Stone on the body and Susan Herring on how men and women discourse differently in cyberspace (1996); to Langdon Winner's passionate 1996 article arguing for the computer profession to get involved with the public in discussing the future. It is particularly valuable that the selections are nearly all of their original length. Some of these writers continue to submit to a journal, *The Information Society*, edited by Rob Kling. The International Federation for Information Processing (IFIP), which is the international body to which all the national computer societies belong, have working groups which have explored these for decades. They are all worth a look, but I pick out Berleur, Clement et al. (1990). A good reader of short selected passages is Dunlop and Kling (1991), including one of my favourite bits of writing by the American essayist Wendell Berry, 'Why I am not going to buy a computer'. It is however ageing in some areas.

Christian introductions include BSR (1999) and Lochhead (1997), the latter a great enthusiast for cyberspace and a participant in experiments in online church. The more general question of religion in society and the impact of the media and information on religious expression is tackled by Lyon (2000) and he finds new initiatives outsides the established religious structures. There seems to be much more Jewish writing than Christian, and I enjoyed discovering a Bible study on Moses in the middle of a book about introducing emotion into computers (Gelernter 1994), and the question of image and idolatry explored in Kochan (1997).

Technology and information

To get a flavour of the technical aspects underlying the information society, read Negroponte (1995), short chapters based on ideas from his columns in the magazine *Wired*. A perspective focused on history rather than possibility is given by the inventor of the Web, but he includes too his ideas for the future – Berners-Lee (1999). A wide-ranging account of the possible future was gathered by the Association for Computing Machinery (ACM) in the USA (Denning and Metcalf 1998). Two books have examined the nature of the interface: Reeves and Nass (1996), in which they describe how people respond to the interface as if it were reality, and Johnson (1997), where he explores the transfer of metaphors, such as the desktop, from reality onto computers and then back into our viewing of reality, for example in the colloquial use of 'windows'.

Information is not as often treated on its own; Borgmann (1999) is interesting in his analysis as to what information is (*intelligence* provided, a *person* is informed by a *sign* about some *thing* in a certain *context*). Brown and Duguid (2000) emphasise how quite intelligent people seem to become so attracted to the data itself that they forget what it took to get it to them, and so make quite unjustified assumptions about the restructuring of the information industry.

Community

Early writing about online communities was initiated by long-term members and researchers: Rheingold (1994) reflects on those experiences, while Hiltz

(1984) analyses and describes what happens to academics and business people in their use of electronic communication. Of more general interest, perhaps, is the recent work on virtual communities, as found in the anthology edited by Jones (1997). In the midst of the overwhelming effect of real words online, Locke (1999) argues that we need to commune, to pass more time in mutual verbal grooming, in order to build community. Turkle (1996) has explored the development of the self and its conceptualisation, and Virilio (1997) the development of 'terminal man' who sits and limits the world to what is accessible through the screen.

Politics

Everard (2000) offers a useful overview of the issues associated with politics; Castells (see above) covers political and governmental aspects of Internet, as does Feenberg. The particular aspect of surveillance is covered in Lyon and Zureik (1996) and a brief and useful philosophical discourse on privacy can be found in Advisory Forum on Human Rights (1984).

Miscellaneous

Technical development is not only done with rationality: in Davis (1998) it is demonstrated that there is a mystical or spiritual basis to much of technical development. Wertheim (1999) claims that cyberspace is another kind of reality similar to the medieval Christian heaven. I have not explicitly considered Artificial Intelligence in this book; if you want an introduction, I suggest Puddefoot (1996) and Emerson and Forbes (1989), both in very accessible styles. For the best description of cyberspace as space, I recommend a novel (Stephenson 1992).

Finally the theological background to thinking on this has emerged from Seerveld (1988) and his persistence in the application of Christ to the very activities of doing whatever we do; Nicholls (1981), for reminding me of the challenges there are in attempting it; Hauerwas (1983) for maintaining the subordinate position of power; Ford (1999) for a review of self; and Williams (2000) for his discussion on how we need to find ways to talk at a societal level.

Bibliography

Advisory Forum on Human Rights, The (1984). *A Study on Privacy: With special reference to computers, technical surveillance and the media*. Belfast, The Irish Council of Churches, Board of Community Affairs.

Barbour, I. (1992). *Ethics in an Age of Technology: The Gifford Lectures*. London, SCM Press.

Berleur, J., Clement, A. et al. (eds) (1990). *The Information Society: Evolving landscapes*. New York, Springer-Verlag.

Berners-Lee, T. (1999). *Weaving the Web*. London, Orion Business.

Board for Social Responsibility (1999). *Cybernauts Awake! Ethical and spiritual implications of computers, information technology and the Internet*. London, Church House Publishing.

Borgmann, A. (1999). *Holding onto Reality: The nature of information at the turn of the millennium*. Chicago, University of Chicago Press.

Brown, J. S. and Duguid, P. (2000). *The Social Life of Information*. Boston, Harvard Business School Press.

Castells, M. (1997). *The Power of Identity*. Oxford, Blackwell.

Davis, E. (1998). *TechnGnosis: Myth, magic and mysticism in the age of information*. London, Serpent's Tail.

de Sola Pool, I. (1977). *The Social Impact of the Telephone*. Cambridge, MA and London, MIT Press.

Denning, P. J. and Metcalf, R. (eds) (1998). *Beyond Calculation: The next fifty years of computing*. New York, Springer Verlag.

Dunlop, C. and Kling, R. (eds) (1991). *Computerization and Controversy: Value conflicts and social choices*. London, Academic Press.

Emerson, A. and Forbes, C. (1989). *The Invasion of the Computer Culture*. Leicester, Inter-Varsity Press.

Everard, J. (2000). *Virtual States: The Internet and the boundaries of the nation-state*. London, Routledge.

Feenberg, A. (1999). *Questioning Technology*. London, Routledge.

Ford, D. F. (1999). *The Self and Salvation*. Cambridge, Cambridge University Press.

Gelernter, D. (1994). *The Muse in the Machine: Computers and creative thought*. London, Fourth Estate.

Graham, G. (1999). *The Internet: A philosophical enquiry*. London, Routledge.

Hauerwas, S. (1983). *The Peaceable Kingdom: A primer in Christian ethics*. Notre Dame, University of Notre Dame Press.

Hawkin, D. J. (1986). *Christ and Modernity: Christian self-understanding in a technological age*. Waterloo, Ontario, Wilfrid Laurier University Press.

Hershock, P. D. (1999). *Reinventing the Wheel: A Buddhist response to the information age*. Albany, State University of New York Press.

Hiltz, S. R. (1984). *Online Communities: A case study of the office of the future*. Norwood, NJ, Ablex Publishing.

Johnson, S. (1997). *Interface Culture: How new technology transforms the way we create and communicate*. London, HarperCollins.

Jones, S. G. (ed.) (1997). *Virtual Culture: Identity and communication in cybersociety*. London, Sage.

Kochan, L. (1997). *Beyond the Graven Image: A Jewish view*. Basingstoke, Macmillan.

Levy, P. (1998). *Becoming Virtual: Reality in the digital age*. New York, Plenum.

Lochhead, D. (1997). *Shifting Realities: Information technologies and the Church*. Geneva, World Council of Churches.

Locke, J. (1999). *The De-Voicing of Society: Why we don't talk to each other anymore*. New York, Simon & Schuster.

Lyon, D. (2000). *Jesus in Disneyland: Religion in postmodern times*. Cambridge, Polity Press.

Lyon, D. and Zureik, E. (eds) (1996). *Computers, Surveillance and Privacy*. London and Minneapolis, University of Minneapolis Press.

Mayer, P. A. (ed.) (1999). *Computer Media and Communication: A reader*. Oxford, Oxford University Press.

Monsma, S. V. (ed.) (1986). *Responsible Technology: A Christian perspective*. Grand Rapids, Michigan, William B. Eerdmans.

Mosco, V. (1989). *The Pay-Per Society: Computers and communication in the information age*. Ontario, Garamond.

Mosco, V. and Wasko, J. (eds) (1988). *The Political Economy and Information*. London, University of Wisconsin Press.

Negroponte, N. (1995). *Being Digital: The road map for survival on the information highway*. London, Hodder & Stoughton.

Nicholls, D. (1981). *Holiness*. London, Darton, Longman & Todd.

Puddefoot, J. C. (1996). *God and the Mind Machine: Computers, artificial intelligence and the human soul*. London, SPCK.

Reeves, B. and Nass, C. (1996). *The Media Equation*. Cambridge, Cambridge University Press.

Rheingold, H. (1994). *The Virtual Community*. London, Secker & Warburg.

Robins, K. and Webster, F. (1999). *Times of the Technoculture: From the information society to the virtual life*. London, Routledge.

Seerveld, C. (1988). *On Being Human: Imaging God in the modern world*. Burlington, Ontario, Welch Publishing Company Inc.

Stephenson, N. (1992). *Snow Crash*. London, Roc/Penguin.

Teich, A. H. (1997). *Technology and the Future*. New York, St Martin's Press.

Tenner, E. (1996). *Why Things Bite back: New technology and the revenge effect*. London, Fourth Estate.

Turkle, S. (1996). *Life on the Screen: Identity in the age of the Internet*. London, Orion.

Virilio, P. (1997). *Open Sky*. London, Verso Books.

Wertheim, M. (1999). *The Pearly Gates of Cyberspace: A history of space from Dante to the Internet*. London, Virago.

Williams, R. (2000). *Lost Icons: Reflections on Cultural Bereavement*. Edinburgh, T. & T. Clark.

Index